A Descriptive History of the Steam Engine

A Descriptive History of the Steam Engine

Robert Stuart

It must be acknowledged that this is the most wonderful of all machines, and that nothing of the work of man approaches so near to animal life. Heat is the principle of its movements; there is in its tubes a circulation like that of the blood in the veins of animals, having valves which open and shut in proper periods; it feeds itself, evacuates such portions of its food as are useless, and draws from its own labours all which is necessary to its own subsistence.—BELIDOR

NONSUCH

First published 1824
Copyright © in this edition Nonsuch Publishing, 2007

Nonsuch Publishing
Cirencester Road, Chalford, Stroud, Gloucestershire, GL6 8PE
www.nonsuch-publishing.com

Nonsuch Publishing is an imprint of NPI Media Group

All rights reserved. No part of this book may be reprinted or
reproduced or utilised in any form or by any electronic, mechanical
or other means, now known or hereafter invented, including
photocopying and recording, or in any information storage or retrieval
system, without the permission in writing from the Publishers.

British Library Cataloguing in Publication Data.
A catalogue record for this book is available from the British Library.

ISBN 978 1 84588 436 9

Typesetting and origination by NPI Media Group
Printed in Great Britain

Contents

Introduction to the Modern Edition	7
Dedication	11
Preface	13
A Descriptive History of the Steam Engine	17
Chronological List of Patents	161
Notes	177

Introduction to the Modern Edition

MAN HAS KNOWN OF THE power of steam for the best part of 3,000 years. Hero the Elder is considered to have written the very first account in his *Spiritulia* of the forcing pump, powered by condensed air. Little useful work was done on the steam engine from then until the 1560s, when Mathesius proposed a steam engine in his *Sarepta*. From then onwards, a new type of engine was developed every thirty or so years, but the real breakthrough came when Bishop Wilkins mentioned in an English publication the possibility of moving a machine by steam power.

Wilkins' theory was taken up by the Marquess of Worcester during the reign of Charles II. He published details of his engine in 1663. Worcester's engine used a separate boiler to make steam, which was forced into another cylinder, and the pressure of the steam would force water out of a jet. Using stop cocks the pressure could be increased or reduced and water added to the boilers.

Sir Samuel Moreland, in 1683, exhibited a steam engine at the French Court, but failed to achieve any encouragement from the French Government. The next major development came from Captain Savery, who managed to make a working steam pump. Savery's engine was relatively powerful and it was obvious that steam could be used to raise water, therefore making one of the first workable steam pumps. Unfortunately for him, his invention was overlooked by those who had most to gain from its application, and he had managed to build and have in operation several steam engines before finally being granted a patent in 1698. In 1696 he published *The Miner's Friend*, which described the engine in reasonable detail, and he also displayed his engine at Hampton Court Palace in front of King William III.

Slowly but surely the development of the steam engine took place, with the next major improvements occurring when Newcomen and Cawley, a blacksmith and glazier respectively, developed the ideas of Papin, in France, and the Marquess of Worcester. Working in Dartmouth, they developed Papin's idea for a cylinder whereby the steam would condense and move a piston to make a working steam engine. With the recently developed safety valve, they had succeeded in making a working and relatively reliable engine. One was installed in 1714 at a coal mine in Griff, Warwickshire. It took a few attempts to get the engine working satisfactorily, but by trial and error, Newcomen succeeded. Working at atmospheric pressure, the engine was more reliable and fuel efficient than Savery's and improvements helped increase these efficiencies.

About 1720, Leupold designed a high-pressure lever engine, while De Moura, a Portugeuse gentleman, further developed Savery's engine. In 1758 Dean Fitzgerald published in *Philosophical Transactions* a method of converting the reciprocating motion of the Atmospheric engine into a rotary motion, utilising large-toothed wheels and smaller ratchet wheels acting in unison. But it was in 1765 that the major development of the steam engine occurred—a development that was to make it the powerhouse of the Industrial Revolution.

Technological progress had been slowly building, and all sorts of things were happening across Britain by the 1760s. Iron works were being built, mechanisation was beginning to take place and the period was to become known as the start of the Industrial Revolution. That revolution was built on steam power and it was the technological improvements made by James Watt that heralded this new age. He realised that the steam cylinder should not be allowed alternately to heat up and cool down as in Newcomen's engine, but be kept heated all of the time. This would lead to an increase in the thermal efficiency of the engine and, as a consequence, an increase in the possible work achieved from it. He accomplished this by the simple expedient of producing a vacuum without cooling the cylinder, and it was his work on a model of Newcomen's engine and discussions with his friend Dr Black that led him to his breakthrough. Installing an engine of his design into a coal mine at Kinneil, near Bo'ness in Scotland, he applied for and won a patent for his engine.

Watt's patent led him into partnership with Mathew Boulton and they started the great company of Boulton & Watt, in Birmingham's Soho district. In 1775 Watt succeeded in gaining an Act of Parliament which extended his patent application for twenty-five years, rather than the standard ten. This was a blow to the development of the steam engine, but it made Boulton and Watt very rich men, as they gained fully a third of the savings in coal on each engine that they made. Of course, they were the only manufacturers of the more efficient engine, so had an effective monopoly on the trade in Britain. Each engine was fitted with a meter and the number of strokes of the piston used to calculate the royalties owed to Boulton and Watt.

Watt's patent effectively stifled development of the steam engine for twenty-five years, but development was rapid afterwards and by the 1830s the steam engine was a common sight in industrial towns and in mines for pumping water and lifting coal, tin and iron ore from the depths.

The steam engine truly changed the world and by the 1830s it was evident that it would power the economic growth of those countries that had embraced it. In 1771 the Frenchman Cugnot had made the world's first steam vehicle and steam engines were slowly finding their way into boats, and by 1804 Richard Trevithick had built the world's first successful steam-railway locomotive. The development of the engine and the multitude of uses it was now put to was rapid; the period from 1800–1830 saw the building of the world's first sea-going steamships and intercity railway, and the increased demand for engines meant a subsequent increase in the numbers of mines for coal to power them and iron to manufacture them. The Industrial Revolution therefore grew exponentially with the increase in new uses for the steam engines.

Much of the development of the steam engine, which was to have such momentous effects, occurred in Britain. It made labour cheaper, it encouraged the growth of the cities, it made mass-production possible, it made long distance travel an everyday event, but, most of all, it created an insatiable demand for products made by steam.

Robert Stuart's *Descriptive History of the Steam Engine* was first published in 1824, at a time when this technological advance was only really gathering pace. He must have been amazed at the advances in technology that were occurring as he wrote what was the first proper history of the steam engine. He had already seen the development of the first railways and the use of steam in boats and ships, but for him the developments that were to take place after he had written his book were perhaps more world-changing than those of the previous 3,000 years. Within fifteen years steam would be king in Britain and, subsequently, the world. Railways had already begun to connect the industrial towns and cities and steamships were beginning to traverse the globe, while the steam engine could be found in almost every industrial concern from cotton mills to coal mines and from ironworks to shipyards. Without the steam engine the industrial development of the world would have been much slower and the world owes a debt of gratitude to those men who developed it. Thankfully, Robert Stuart was around to compile his history and, without those men listed in his Chronological List of Patents at the end of this volume, much of our present technology would not exist.

Campbell McCutcheon
September 2007

To

Doctor George Birkbeck

President of the London Mechanics' Institution, Patron of the Glasgow Mechanics' Institution, President of the Meteorological Society, &c. &c., who, while Professor of Natural Philosophy in the College founded by Professor Anderson, in the City of Glasgow, first drew together an audience exclusively composed of artisans, to explain to them in a manner adapted to their comprehension, those principles of philosophy on which their manual operations depended, with the avowed object by enlarging their minds to raise their rank in society, and by adding scientific precision to mechanical skill to make their labour lighter in itself, and more efficient for their country: this attempt to present to working mechanics, in a plain manner and cheap form, a descriptive history of the progress of improvement made in the steam engine is respectfully dedicated by the author.

Preface

THE WONDERFUL INCREASE OF MACHINERY, of all kinds, that has taken place in England during the last thirty years, having been mainly produced by those improvements made on the Steam Engine which led to its general introduction as a first mover, has conferred great interest and importance on every circumstance connected with the construction and history of that magnificent mechanism.

Several treatises of great merit, describing particular engines, have been long before the public; but in none of them has the general history of improvement formed any part of the plan of their authors.

With the view of supplying this want, the following compilation was undertaken. It was considered too, that the meritorious classes particularly, who are engaged in the construction of machinery, and in directing its operations when applied to manufactures, and who, as best knowing its importance, must feel a greater interest in the history of Steam mechanism, would receive with indulgence any attempt to place this information within the reach of their attainment; for among those whose time is nearly filled up by manual exertions to make "provision for the day that is passing over them," a want of leisure must prevent the greater number from consulting the numerous volumes in which the facts of its history lie scattered—even supposing that their means permitted their incurring the great expense at which this information could alone be purchased.

In a plan, therefore, which embraced so wide a field, and which was yet to be brought within the reach of mechanics in general, the distinguishing features only of each apparatus could in most cases be noticed; the details of many engines are therefore omitted, where they may be similar to any one previously described, or where their office and action may be easily understood by what has gone before: this omission has also been sometimes made in the Figures. Practical men, however, will not object to a very plain matter not being repeated, and general readers, (should any light on this volume,) by referring to the Figures, it is hoped, will find no difficulty in filling up the description: but to both classes of readers an inspection of a Figure will give a clearer idea of the action of machines, (nothing else can explain their construction,) than the most lengthened description;—a line of engraving is worth pages of letter-press in the explanation of machinery. It was under this impression that so great a number of Figures was thought necessary, as are given in this volume. Some of these Figures have been selected with a view to give an idea of the appearance of the Steam Engine at various times; and others to explain its action without any pretensions beyond that of mere explanatory diagrams.

The same reasons which prompted the omission of descriptions of some of the details, have applied to the exclusion of merely theoretic disquisition or inference from these pages. This will not probably be objected to, since it so happens that the little which has been done by learned men on this subject is of no practical 'mark or likelihood.' Twenty years ago Hornblower remarked, "that the most vulgar stoker may turn up his nose at the acutest mathematician in the world, for, (in the action and construction of Steam Engines,) there are cases in which the higher powers of the human mind must bend to mere mechanical instinct;" and the observation applies with greater force now than it did then.

We know not, therefore, how the remark has originated, or what "philosopher" first claimed for theoretic men any part of the honour of being instrumental, even indirectly, in the perfecting of the Steam Engine; or who gave currency to the phrase of its "invention being one of the noblest gifts that *science* ever made to mankind!" The fact is, that science, or scientific men, never had anything to do in the matter. It was a toy in the hands of all the philosophers who preceded Savery, and it again must become a toy before the speculations of Bossut, the ablest and latest of the philosophers who have written on the subject, can be made to bear upon it. Indeed, there is no machine or mechanism in which the little that theorists have done is more useless. The honour of bringing it to its present state of perfection, therefore, belongs to a different and more useful class. It arose, was improved and perfected by working mechanics—and by them only; for tradition has preserved to us the fact of Savery having begun life as a working miner;—Newcomen was a blacksmith, and his partner Cawley a glazier;—Don Ricardo Trevithick was also an operative mechanic; and so was the illustrious Watt, when he began, and after he had made his grand improvements.

A Descriptive History of the Steam Engine

Fig. 1

A Descriptive History of the Steam Engine

Although the elastic power of the vapour of water must have been familiar to man from the earliest period of his history, the first recorded observation of the fact, and the application of Steam to generate motion, appear to have been made by a Greek mechanic, about one hundred and thirty years before the Christian era.

Hero the Elder, who flourished at Alexandria in the reign of Ptolemy Philadelphus, was eminently distinguished in that age and region of refinement, not only for the extent of his attainments in the learning of the time, but also for the number and ingenuity of his mechanical inventions. In one of his books, he deduced all the laws of what are called the mechanical powers from the properties of the lever. His *Spiritalia*, or Pneumatica, contains the first account of the forcing pump: of a fountain, still known by his name, in which water is elevated in a jet by the elasticity of condensed air. Among other contrivances in the same treatise, he describes two machines of his invention; in one of which a rotatory motion is produced by the emission of heated air; and a similar movement is imparted to the other by the reaction of vapour rising from boiling water.

A pipe, *a,* is directed by Hero to be inserted under the hearth of an altar, (Figure 1,) on which a fire is burning. This pipe, placed in a vertical position, is moveable on a pivot, *b,* resting on the base of the altar. Two other pipes, *c, d,* of smaller diameter, proceed from the vertical one in a horizontal direction, having their extremities, *e, f,* open, and turned upwards. A base, or drum, *g,* is attached to the pipes, on which are placed small figures in various attitudes. The air at the upper extremity of the vertical pipe being heated by contact with the under side of the altar hearth, is expanded, and descends into the pipe, and proceeding along the horizontal arms, is expelled at their orifices, *e,*

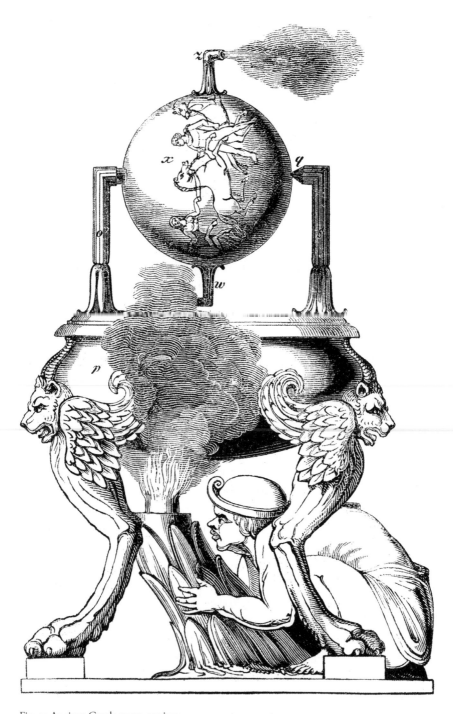

Fig. 2: Ancient Greek steam engine

f. This causes them to revolve round the pivot *b*, so that the figures which are placed on the base *g*, are carried round with it, and appear "to lead the dance, as if they were animated beings."

It is scarcely necessary to notice the identity of this elegant apparatus with that of Barker's mill; and that the rotatory motion would be produced, as stated by Hero, though not by the *emission* of *warm*, but through the *admission of cold* air at the orifices in the horizontal arms, in consequence of the rarefaction at the upper end of the vertical pipe under the hearth of the altar.

The second machine is constructed on a similar principle;—a globe moved on a pivot, by means of steam conducted into it from a boiling caldron.

The caldron or heated vase, *p*, in Figure 2, is to be closely covered with a lid into which a pipe, *o*, is inserted atone side of its circumference. This pipe, after rising vertically for a short distance, is bent at right angles. On its horizontal end is placed a small globe, *w*, kept in its position by a pipe, *s*, also bent at right angles and fixed to the lid opposite to *o*, but terminating in a pivot, *q*, on which the little globe revolves. This globe is furnished with two small pipes, *z, w*, bent at their extremities and open. The steam from the boiling water in *p*, thing through the pipe *o*, is admitted at *s* into the globe; and issuing through the bent tubes *z, w*, causes the sphere to revolve as if it were "actuated from within by a spirit."[1]

This simple and effective apparatus, though described but as a philosophical toy, is curious, as being the primitive mode in which steam was applied to produce motion, and as conferring on Hero the honour of having invented and constructed the First Steam Engine.[2]

No other notice of Steam as a first mover occurs in the works of ancient authors; nor in modern writers until about the year 1563; when one Mathesius, in a volume of sermons, entitled *Sarepta*, hints at the possibility of constructing an apparatus similar in its operation and properties to those of the modern Steam Engine.[3] About thirty years after this period, what is called a "*Whirling Oelipile*" (shown in Figure 3,) is described in a book printed at Leipsig[4], wherein it is stated to be exceedingly well adapted to the purpose of turning the spit for the cook. And among other economical reasons urged for its adoption, is that "it eats nothing, and gives withal an assurance to those partaking of the feast, whose suspicious natures nurse queasy appetites, that the haunch has not been pawed by the turnspit (in the absence of the housewife's eye), for the pleasure of licking his unclean fingers."

A small quantity of water is introduced into the globe, *x,* (in Figure 3), which is rarefied into steam by a fire made under it. The vapour issues at the necks *a* and *b*, and by its reaction a continuous motion is generated.

Solomon De Caus[5], an eminent French engineer and mathematician, in 1624, describes an engine acting by the elasticity of Steam. It consists of a spherical vessel, *m, w,* placed over a fire. This vessel has two apertures; into one is fitted a pipe, *n,* which has a stop-cock, *o,* and funnel *a*: this supplies water to the boiler. The other orifice has a pipe, *x,* which descends through the water, until it nearly touches the bottom of the vessel, and rises to some convenient height above it. When the water, *c,* becomes heated, De Caus says, that the increased bulk of the vapour forces the water up the pipe *x,* which issues in a jet at *b.* De Caus was also acquainted with the fact that Steam could be condensed into its own weight of water; but he appears to have been ignorant of any mode of applying this property to aid the effect of his fountain.

The first person in modern times who applied the expansive power of Steam on any scale to a useful practical purpose was Giovanni Branca, an eminent Italian mathematician who resided at Rome in the beginning of the seventeenth century. His contrivance was an *oelipile*, from which steam issued, upon a wheel formed with float-boards or vanes like a water-wheel or wind-mill, and thus produced a rotatory movement. This wheel, by some intermediate mechanism, gave motion to the stampers of a mill for pounding drugs. Our Fifth Figure is copied from that given by Branca to explain his

Fig. 3: German engine

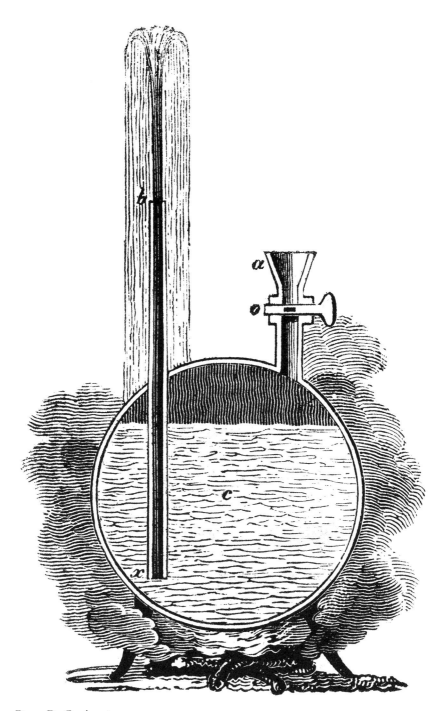

Fig. 4: De Caus' engine

Fig. 5: Branca's engine

invention; but it must be considered only as an ornamental and picturesque illustration of the principle by which he produced the moving power in his stamping-mill; not as a view of any part of the machinery which was actually constructed. *a*, is a boiler in the shape of a Negro's head. *b*, a pipe proceeding from it, which conducts the steam upon the vanes or boards of a wheel, *x*. Other wheels, *e, f,* are attached in the usual manner to communicate the motion in the required direction.[6]

It is on account of this contrivance that Branca is considered by his countrymen to be the inventor of the Steam Engine; and even in a recent English work[7] on the subject, he is allowed the merit of a *first idea*. To this he certainly has no claim; neither can his engine be compared with Hero's for its ingenuity, nor to De Caus's for its efficiency. Besides, long before this period, the same mechanism was described, by Cardan as moved by the "vapour from fire." The mere substitution of steam by the Italian philosopher is not so original or important, as to give to the transition the rank of an invention. Branca was, however, a man of much ingenuity, and many of his machines are highly creditable to his abilities as a scientific mechanic.

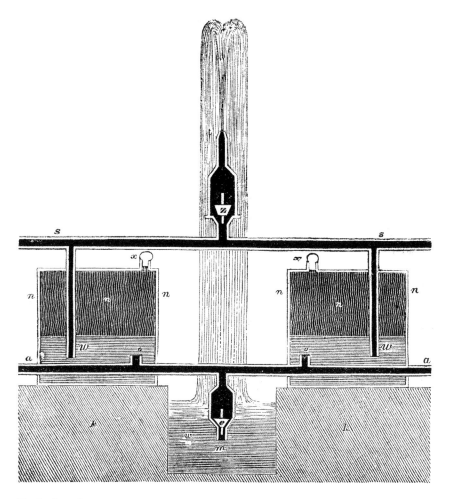

Fig. 6: air engine

The elasticity of the vapour of water, which had long been known to philosophers, but to them only, had now become familiar to water-work artists; and in their hands it was applied in a variety of ways to their favourite problem of raising water above its level in jets and fountains. Without vouching for the great effects said to be produced by these machines, we will describe two, as necessary to give a clear notion of the value of these conceits, and as specimens of the ingenious absurdities, which, under the name of *Air Engines*, were recommended even by experienced engineers about this period. The machines themselves, under another form, are to be found in the *Spiritalia*. The book from which they are extracted in their present shape was one of some reputation in its day, and many years after its publication it was thought worthy of being translated into English. The translation went through two editions.[8]

Figure 6 represents a *"Very subtil engine to raise Water by means of the Sun,"* which, according to its inventor, "hath great effect in hot places, as in Spain and Italy; because in those places the sun shines almost always with great heat, especially in the summer." And they are to be constructed after this manner:

> You must have four vessels of copper, (in our engraving we have only shown two, *n, n,*) well soldered round about, each of which must be about a foot square, and eight or nine inches high. A pipe, *s s;* is placed on each vessel, having other pipes, *w, w,* attached to it, reaching almost to its bottom. A sucker, or valve, *z,* is placed in the middle of the pipes, made and placed so that, when the water springs out of the vessels, it may open, and being gone forth, may shut again. You must also have another pipe, *a, a,* with small pipes, *o, o,* rising upwards at the bottom of these vessels, and also a sucker (or valve *e*) to the end of which there is a pipe *m,* which descends into the water in the cistern *r.* To the copper vessels *n, n,* there shall also be vent-holes *x, x.* So placing the engine in a place where the sun may shine upon it, pour water into the vessels *n, n,* by the holes *x,* to about a third part of their content; the air with which they were previously filled will pass out by the passages. Afterwards you must stop all these passages; and then the sun shining upon the engine shall make an expression because of the heat, which causeth the water to rise from all the vessels by the pipes *w, w,* and pass forth by the valve *z*; and when there shall be a great quantity of water run forth by the violence of the heat of the sun, then the valve *z* shall return; and after the heat of the day is passed, and the night shall come, the vessels, to *shun vacuity*, shall draw up the water of the cistern by the pipe *a, o, m,* and shall fill the vessels as they were before; so long as there is any water in the cistern.

But, continues the translator, if you desire to raise the water five or six feet high, "the foregoing engine cannot raise it if the sun does not shine with great violence." "To increase the force of the Sun," he proposes to improve the "Subtil Engine" by forming the copper vessels, as shown in Figure 7 where A, A, are burning glasses well fitted into the sides of the copper vessels. These glasses are placed towards the south and west, "that the sun shining upon them may assemble the rays of the sun within the said vessels, which will cause a great heat to the water, and by that means make it spring forth in great abundance." X is the copper vessel communicating by the pipe O, with

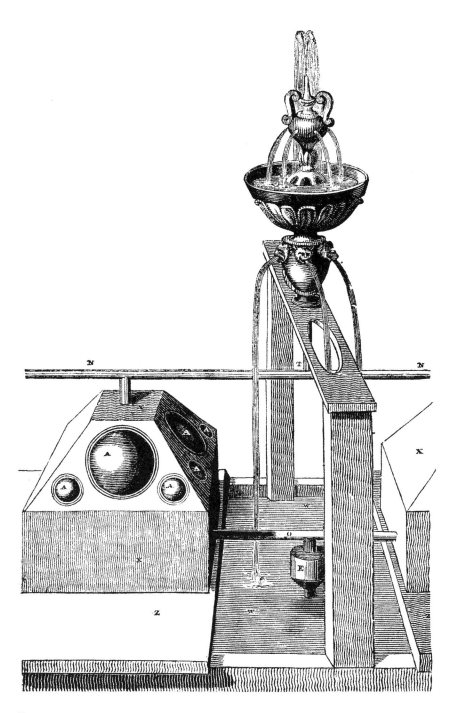

Fig. 7: air engine

the pipe and valve E in the cistern W. N, N, are pipes in which the water rises through T, into the fountain placed over it. Z is a floor over the cistern for supporting the copper vessels.

The learned and ingenious Bishop Wilkins is the first English author who mentions the possibility of moving a machine by the elastic force of Steam. Speaking of giving motion by wind or air, "Something of this nature," says the Bishop, "are the *oelipiles*, which are concave vessels, consisting of some such material as may endure the fire, having a small hole, at which they are filled with water; and out of which, when the vessels are heated, the air doth issue with a strong and lasting violence. These are frequently used for exciting and contracting of heat in the melting of glasses or metals. They may also be contrived for sundry other pleasant uses; as for moving of sails in a chimney-corner; the motion of which sails may be applied to the turning of a spit or the like."[9]

This is Cardan's application, and Branca's engine. The whole passage is curious; for, from the manner in which the moving of sails is mentioned by the Bishop, the contrivance would appear to have been used in England at this period. We have now no means of ascertaining whether it was known in England, or to the Bishop, before the date of Branca's book; or whether it was copied from the design of that ingenious Italian.

But among all those whose names are associated with the history of the Steam Engine in its infant stages, a Marquis of Worcester, who lived in the reign of Charles II, is by far the most celebrated. This distinction is the more extraordinary when we take into consideration the neglect with which his extravagant pretensions to the merit of discoveries were treated in his life time; the studied brevity, vagueness, and obscurity of those descriptions of his contrivances, on which he rested his claim of merit and his demand for encouragement; and the long oblivion in which they lay after his death—that brilliant homage which in our time is awarded to his mechanical genius, as far transcending his real merit, as it was unjustly under-rated by the total neglect of his contemporaries.

His claims to invention, after all, resting solely on his *own* account of the *uses* and *wonderful properties* of his contrivances, the confidence which is due to those statements can only be fairly estimated by a reference to the general probity of the Marquis's character: a test, which, if Lord Orford's sketch bears any resemblance to the original, will deter us from placing any reliance whatever on the unverified explanations in the *Century of Inventions*.[10]

This book, on the merits of which its author was loud in his demand for national patronage, called by Walpole with much truth "an amazing piece of folly," was published by the Marquis himself, under the title of "A

Century of the Names and Scantlings of such Inventions as at present I can call to mind to have tried and perfected, (my former notes being lost,) 1663."[11] It had two dedications: the first to King Charles II; and the second to both Houses of Parliament. In the second address he affirms having, in the presence of the King, performed many of the "feats" mentioned in his pamphlet. The sixty-eighth article or description in this book, is that on which is generally rested his claim to the honour of having invented the Steam Engine. It is in these words:

> I have invented an admirable and forcible way to draw up water by fire; not by drawing or sucking it upwards, for that must be, as the philosopher terms it, *infra sphoeram activitatis*, which is but at such a distance; but this way hath no bounder, if the vessel be strong enough. For I have taken a piece of a whole cannon, whereof the end was burst, and filled it three quarters full of water, stopping, and screwing up the broken end, as also the touch-hole; and making a constant fire under it: within twenty-four hours it burst, and made a great crack. So that having a way to make my vessels, so that they are strengthened by the force within them, and the one to fill after the other, I have seen the water run like a constant stream forty feet high. One vessel of water rarefied by fire, driveth up forty of cold water; and a man that tends the work has but to turn two cocks; that one vessel of water being consumed, another begins to force and refill with cold water, and so successively; the fire being tended and kept constant, which the self same person may likewise abundantly perform in the interim between the necessity of turning said cocks.[12]

This account, according to Professor Robison, "although by no means fit to give us any *distinct notions* of the *structure and operation* of his engine, is exact *as far as it goes*, agreeing precisely with what we know of the subject." But the Professor afterwards adds, "that the account in the 'Century of Inventions' could instruct no person who was not sufficiently acquainted with the properties of Steam, to be able to *invent the machine himself*." And yet in the same treatise the Doctor says the Steam Engine "was beyond all doubt invented by the Marquis of Worcester!" Hero's, De Caus's, and Branca's engines were unknown to Dr Robison: not so the overwhelming quackery of the Marquis of Worcester[13], and the absurd extravagance of his pretensions.

With the slight alteration of substituting and pipe in the centre, for a pipe placed at each extremity in Mr Millington's arrangement, our Eighth Figure represents an apparatus which that ingenious mechanic has designed from the account in the *Century of Inventions*. The failure of a person of so much judgment and experience in the combination of Steam machinery to produce an engine fulfilling the conditions of the enigma, *and no more*, gives us a pretty clear notion of the value of the claim to discovery, by showing the impossibility of the problem.

The two spherical vessels *a, o,* in Figure 8, have two pipes, *d, f,* proceeding from them, and inserted into a boiler, *g.* These pipes have two stop-cocks, *z, w,* which shut off the communication between the boiler and the vessels. From another part of the vessels proceed two other pipes, having valves at *s, x,* opening outwards, and terminating in a single pipe, *e.* The spherical vessels have each another valve opening inwards, and a very short pipe, *n, v*; the pipe *n, e,* rises forty feet, and terminates in the reservoir *u. b,* is a section of the fire-grate, under the boiler *c*; *t,* the door of the fire-place; *l,* the brick-work; *g,* the ash-pit; and *h,* is a reservoir of water in which the vessels *o, a,* are placed, and which is to be elevated into the cistern *u.*

If we now suppose a sufficient quantity of steam to be generated in the boiler *c,* from the water *g,* and the stop-cock *z,* opened so as to allow a free communication between the boiler and the vessels in the reservoir, the steam will descend in the pipe *d,* (the pipes and vessels being made or cased with some material, to prevent the condensation of the steam by the water in the reservoir,) into the vessel *a,* and will expel all the water or air which it may contain through the valve *s,* into the pipe *e,* which will deliver it into the reservoir *u.* The cock *z* is now to be shut, and the valve *v,* being freed from the pressure of the elastic vapour, will be forced inwards by the gravity of the water in the reservoir, which will speedily fill the vessel *a.* But when the cock *z,* is shut, the opposite one, *w,* is opened, and the steam from the boiler raises the water which may be contained in *o,* up the pipe closing in this operation the valve *s.* When the vessel *o,* is filled with steam, the cock *w,* is shut, and the water in the reservoir rushes into *o,* as it did into *a,* and fills it. The cock *z,* is now opened and the steam again expels the water from the vessel *a;* and so on successively, so long as steam is produced in the boiler, and the cocks *z, w,* are opened and shut alternately.

Mr Millington remarks that this engine agrees so far with the Marquis of Worcester's description, where he says that "a man has but to turn two cocks, and that one vessel of water being consumed, another begins to force and refill." He also observes that the condensation of the steam opens and shuts the valves, and fills the vessels, but that this use of the vacuum is part of an invention to which the Marquis has no claim, his Lordship expressly stating, that "the water *is not* raised by drawing or sucking it upwards." The "force and refill" in the original account would almost lead to a supposition that

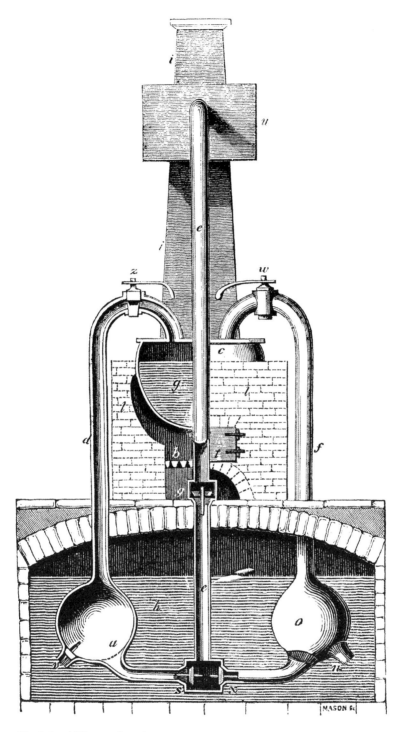

Fig. 8: Lord Worcester's engine

these operations were going on at the same moment, in the same vessel. The arrangement of pipes and cocks and valves is also gratuitous.

The "admirable method of drawing up water by fire appears to have been the favourite project of the noble inventor; for he afterwards devoted a separate book to an enumeration of its extraordinary uses and powers, under the title of an "Exact and True Definition of the most stupendous Water commanding Engine, invented by the Right Honourable (and deservedly to be praised and admired) Edward Somerset, Lord Marquis of Worcester; and, by his Lordship himself presented to his most excellent Majesty: Charles the Second; our most gracious Sovereign." This "Exact and, True Definition" is a quarto pamphlet, of twenty-two pages; but, instead of *a definition,* it contains only an enumeration of the *marvellous uses* of his invention, as vaguely and obscurely written as those in the *Century of Inventions.* The rest of the pamphlet is filled up with an Act of Parliament, allowing him the monopoly of such an engine, and reserving the tenth part of the profits to the King, with four wretched verses of *his own*, in commendation of his invention; with the "*Exigi monumentum*" of Horace, and the "*Barbara Pyramidam sileat*," of Martial. Some Latin and English verses, panegyrizing the noble inventor, written by James Rollock, an old dependant of his Lordship, complete the volume.

If the Marquis ever made an experiment on the elasticity of steam, (for the bursting of the cannon is a truly questionable one,) or if he ever attempted to carry his project into execution by constructing an engine, all records of his experiment and apparatus are lost. The more probable opinion is that he never made either the one or the other; and this surmise is almost strengthened into a certainty from a clause in the Act of Parliament granting him the privilege of monopoly; for it is there expressly stated, (and the statement is a proof that he proceeding was an unusual one,) that the patent was secured to the Marquis, "on his *simple* affirmation of his having made the discovery." It were almost superfluous to suggest the improbability of this statement having been made, if he could have referred to the evidence of an engine or of an experiment.

Tradition has preserved nothing on this subject, except an anecdote[14] of the Marquis's attention having been first drawn to the amazing force of steam, from observing the rising of the lid of a vessel which was employed in some culinary operation in his chamber, when he was confined in the Tower of London. We should, however, pay a sorry compliment to that "learning and industry" for which this nobleman is almost as much celebrated as for his ingenuity, were we to suppose him ignorant of Branca's book; or of De Caus's inventions, published so near his time, in a country where he resided many years, and where the volume which contained them was a common book of reference to the philosopher and mechanic, even to a period long after that of the publication of the *Century of Inventions*. And, lastly, we must suppose the Marquis acquainted with his contemporary Dr Wilkins's project of the *Oelipile*.

It must, however, be acknowledged that although his merit is at an immeasurable distance from that of the inventor, the Marquis of Worcester is entitled to some mention as the probable projector of an improvement in Steam Engine Apparatus. He may have imagined the advantages which would arise from adding another vessel to De Caus's contrivance, (shown in Figure 4,) in which the steam might be generated, and then admitted into a second vessel filled with cold water. But even this improvement (and it is an important one) can only be allowed to his Lordship after making an emendation, or correcting an error, in his account in the *Century of Inventions*. Instead of reading "to force and refill with cold water," it would be necessary to say, "to force and empty of cold water."[15] Yet the manuscript in the British Museum agrees, in the usual reading, with the printed copies.

From the opinion we have expressed of its being practically impossible to produce an apparatus fulfilling *all* the conditions of the description in the *Century of Inventions*, without introducing parts which are unquestionably due to the inventive genius of other mechanics, it is with great diffidence we propose the apparatus represented in Figure 9, as being nearer to the description than that shown in Figure 8, in so far as *not using the pressure of the atmosphere*, which the Marquis states a *principle of his engine*.

But, after all, it is impossible to decide from Lord Worcester's description whether *two boilers* are meant, and *one receiving vessel*;—or *two vessels*, and *one boiler*? Or only two vessels, like De Caus's, probably having each an eduction-pipe and the proper cocks, to produce a continuity in the stream of water?—as in Figure 9 where the dotted lines rising from the vessel *a*, would then represent this pipe, the fire being made under each boiler alternately. And, when mention is made of one vessel of water rarefied by fire, driving up forty of cold water, should not this be understood as the proportion of water which would he converted into steam, in order to raise the remaining portion in the *same vessel*, forty feet high? An apparatus designed on this supposition, would satisfy the description nearer than any.

Having so wide a choice, we will give the Marquis credit for an apparatus for generating the steam in a separate boiler. *a*, will therefore be that boiler; *c*, a pipe (having a stop-cock *d*,) connecting the boiler with the cold water vessel *e*, from which proceeds the eduction-pipe *f*; *g*, a pipe and funnel, to supply the boiler with water; *h*, a similar pipe-cock to supply the vessel *e*, with cold water, connected with a cistern, from which the water is to be raised; *i*, is a stop-cock on this pipe; *k*, a valve to prevent the return of the water which may be in the upper part of the pipe *f*.

When the steam in the boiler *a*, is allowed to enter the cold water vessel *e*, by turning the cock *d*, the water is raised in a jet through *f*, until the vessel *e*, is emptied. When this is the case, the cock *d*, is shut, and *i*, is opened, and the vessel *e*, is again filled with cold water. The cock *i*, is then shut, and the stop-cock *d*, is opened; and the steam from the boiler pressing on the surface of the water in *e*, forces it up the pipe *f*. When this is emptied the same operation is repeated, and so on successively. So that here the condition of the alternate opening and shutting two cocks is fulfilled; also the forcing and refilling of the vessels; and one vessel of water rarefied by fire would elevate double the quantity of that stated by the Marquis.

After the death of the late Doctor Robison of Edinburgh, there was a "List of Dr Hooke's Inventions" found among the Professor's papers, which contained the following memorandum: "1678, proposed a Steam Engine on Newcomen's principle:"—a note, it is to be regretted, that the biographer[16] has allowed to remain without any comment; it being probable, from his known erudition, that he could have supplied the quotation of authority from Hooke's writings. It would also have been interesting to have ascertained whether this memorandum was made by Dr Robison before or after his having written the excellent account of the Steam Engine in the *Encyclopedia Britannica*. The project was either unknown to the Professor, at the period of publishing the article alluded to; or it was rejected on account of its questionable character. In the edition of that account,[17] lately printed, no notice whatever is taken of Hooke's idea. The proposal has never presented itself to us in our perusal of the works of this most philosophical of modern mechanics.

No farther public attempt to raise water by steam (for hitherto, with one exception, this appears to have been the grand problem for solution,) was made until some years after the Marquis of Worcester's death; when, in 1682, we find Sir Samuel Moreland, at Paris, endeavouring to obtain the patronage of the French Government towards a scheme which he claims as his own, for raising water by the force of steam. In 1683 he exhibited his invention to the French King at St Germain's; but he appears to have been unsuccessful in his application to that Court for encouragement. No description of his apparatus, or of the principle or mode of its action, is known to be in existence. Fortunately, however, the results of some experiments made by Sir Samuel

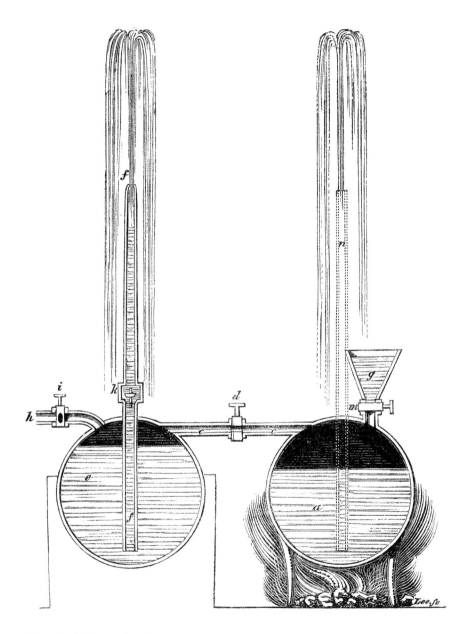

Fig. 9: Lord Worcester's engine

upon the elasticity of steam are still preserved among the manuscripts in the British Museum.[18]

Sir Samuel's account is in the French language, written on vellum, in a very beautiful hand, and highly ornamented. The volume is stated on its title-page to have been presented to the King of France by Sir Samuel Moreland, Master of Mechanics to the King of England. It consists of twenty-two pages only, and purports to be an account of various machines for raising water. The part relating to the Steam Engine does not quite occupy four pages. Considering the period in which this portion of it was written, it is a very remarkable performance.

"Water being evaporated," says Sir Samuel, "by the force of fire, its vapour occupies a much larger space (about two thousand times) than it occupied before. And its power is so prodigious, as, if it were closely confined, to burst a cannon; but being governed upon statical principles, and by science reduced to measure, weight, and balance, it then bears itself quietly under the harness, (like good horses,) and becomes of great use to mankind, particularly for the raising of water, according to the following table, which shows the number of pounds (French) weight which may be raised to the height of six inches eighteen hundred times a minute, in cylinders about half filled with water; as well as the different diameters and depths of the cylinders.

CYLINDERS

Diameter in Feet	Length in Feet	Pounds to be raised
1	2	15
2	4	120
3	6	405
4	8	960
5	10	1875
6	12	3240

Number of Cylinders, having a diameter of 6 Feet, and a length of 12 Feet	Required to raise the following numbers of Pounds weight of Water
1	3,240
2	6,480
3	9,720
4	12,960
5	16,200

6	19,440
7	22,680
8	25,920
9	29,160
10	32,400
20	64,800
30	97,200
40	129,600
50	162,000
60	194,400
70	226,000
80	259,000
90	291,610

Sir Samuel's experiments must have been made with considerable care; and it is highly honourable to his mechanical accuracy that his estimate of the *rate of expansion* of water, when converted into steam, should coincide with that which is given by an experienced engineer in the most recent book on the subject, as an approximation to be relied on in practice. Desaguliers long afterwards stated this expansion at 14,000 times! His estimate, remaining unquestioned for more than half a century, is to be found in all books published prior to about 1800; when Professor Robison gave Mr Watt's experiments, in which it was made between 18 and 1900 times the volume of water which produced it. The mode, too, of stating the quantity of water, which may be raised a certain height a given number of times in a minute, is still the modern method of estimating the power of Steam Engines. It is an obvious remark, that Sir Samuel must have been acquainted with the description in the *Century of Inventions*, from quoting the Marquis of Worcester's improbable experiment as an illustration of the force of vapour. Lord Worcester had been dead several years before the date of Sir Samuel's exhibition of his apparatus; and whatever was its arrangement or principle, he is entitled to the merit of being the first accurate experimenter on the elastic force of steam.[19]

It has been remarked as a curious circumstance in the history of the Steam Engine, that almost every one who made an improvement, either in its construction or application, laid claim to the exclusive merit of having invented the engine. Dr Denys Papin, a native of Blois, a man of great ingenuity, and of considerable acquirements as a philosopher, is considered by his countrymen to be the true inventor of the Steam Engine: a claim strongly contested by some English authors of eminence who have written on the subject,—but on grounds which appear to have been taken from very erroneous and prejudiced statements. It is due to Papin to state that no one, whose labours have produced so many important results, has in his writings shown so little of the vanity and absurd enthusiasm proverbially characteristic of an inventor.

Papin's *first* project[20]—and it is necessary to keep this in remembrance—was to procure a first power by an *air-pump*. This scheme he announced as a means of enabling him to transmit to considerable distances the action of a mill, by means of pipes. The cylinders of air-pumps at one extremity were made to communicate by pipes, with equal cylinders placed at the other and distant end, which by some intermediate mechanism were there connected to the piston-rods of the pumps of a mine. When the mill was put in motion, (by a fall of water, for example,) the pistons of the cylinders to which it was attached were elevated or depressed, as it might be, and the pistons at the other distant extremity of the pipe had an opposite upward or downward movement. When the pistons, for instance, attached to the mill were elevated, the others were depressed, and so on successively. This project failed even on the small scale of an experiment, from the prodigious resistance of the piston, even when one cylinder was attached to the other cylinder by a pipe of only a few inches in length, and the slowness with which the motion could be communicated. This, Papin endeavoured to obviate, by employing some other means of making a *vacuum* under the cylinder, than that of *pumping out the air* by the mill. In 1688, he described an improvement[21] of displacing the air in the cylinder under the piston, by exploding *gunpowder*: but here again it was found that the power was trifling, for without endangering his apparatus he never could make the gaseous product in the under part of his cylinder so entirely fill its capacity as to have no air under the piston. It was now pointed out to Papin, by Dr Hooke and others, that however perfect the exhaustion might be in the *first* and *second* projects, if the pipe of communication were of any length, the compressibility of the air (keeping its friction out of the question) was so great, that unless his cylinder was enormously long, (or the *stroke* of his piston, as it is technically termed,) the motion of the piston at the other end would be imperceptible. However, in 1690, when he published in a separate form his description of this mode of using gunpowder, the idea of transmitting motion to a great distance was perfected by his forming a *vacuum* not only under the piston, but in the *pipe of communication*. Still this scheme, although ingenious, Papin was aware was nearly impracticable, from the great difficulty of abstracting the air by the air-pump or gunpowder; and, among other auxiliary methods which he suggested for obviating this imperfection, is the one of employing *steam* for *forming the vacuum under the piston;*

and also for raising that piston by its elasticity. In this paper he shows that in a little water, changed into steam by means of fire, we can have an elastic power like air; but that it totally disappears when chilled, and changes into water; by which means he perceived that he could contrive a machine in such a manner, that with a small fire he would be able, at a trifling expense, to have a perfect vacuum, which, he admitted, could not be obtained by gunpowder.

In a collection of letters describing some of his inventions, this machine is the subject of his fourth communication. After noticing the difficulty of making a vacuum by gunpowder—"Where there may not be the conveniency of a near river, to play the aforesaid engine," (that of 1685,) he proposes alternately "turning a small surface of water into vapour by fire, applied to the bottom of the cylinder that contains it; *which vapour forces up the plug* (or piston*) in the cylinder to a considerable height, and which, as the vapour condenses*, (as the water cools when taken from he fire,) *descends again by air's pressure, and is applied to raise water out of the mine.*"[22]

This was a happy thought: and had Papin persevered to make the experiment, he would, beyond all question, have produced the *Atmospheric Engine*.[23] And although we cannot refuse to this ingenious man the great honour due to having given the *first idea*—of having laid the foundation of the splendid mechanism of the "Lever Engine," the merit of putting the scheme into practice is as certainly due to another. And it so happened that Papin's claim to the actual construction of a Steam Engine on another principle (and which appears to have withdrawn his attention from the prosecution of his first project) cannot in fairness be allowed to him.

The increasing depth of our mines, and the enormous expense even of the inadequate means which were then possessed of draining them, had become a matter of concern, not only to those more immediately interested in that species of property, but to the nation. The attention of mechanics, thus keenly directed to the subject, was bent rather to devise modes of improving the machinery then in use, by diminishing its friction and in its better arrangement, than in the production of a more efficient and economical power. The history of the mines at this period offers but a series of failures on the part of projectors, and of complaint of those proprietors, who, induced by the urgency of the want, or the cupidity of speculation, had been induced to embark in the expense of making the experiments. Every miscarriage thus added to the obstacles which at all times impede the introduction of improvement; and the abortive attempts of ignorant or designing men were urged as reasons for disregarding the inventions of more honourable and meritorious individuals.

It was at this period that Captain Savery, a seafaring gentleman, offered an engine of his invention to the notice of the Mining Adventurers; "which shewed as much ingenuity, depth of thought, and mechanical skill, as ever discovered itself in any design of this nature:"[24] but, from the repeated failure of other schemes, more specious, and big with promise, Savery's account of the simplicity and power of his apparatus drew but little observation from those more exclusively interested in its adoption. Years after his engine had been invented, and found to be practically efficient, we find Savery, who had made but trifling progress in extending its use, appealing to the actual performance of his engine, to combat the prejudices operating against it as being but a *Project!* "I am not fond," says this great mechanic; "of lying under the *scandal* of a *bare projector*; and therefore present you here with a draught of my machine, (see Figure 10,) and lay before you the uses of it; and leave it to your consideration whether it be worth your while to make use of it or no. I can easily give grains of allowance for your *suspicions*, because I know very well what miscarriages there have been by people ignorant of what they pretend to. These, I know, have been so frequent, so fair and so *promising at first;* but so short of performing what they pretend to, that your prudence and discretion will not now suffer you to believe anything without a demonstration—your appetites to new inventions of this nature having been baulkt too often: yet, after all, I must beg you not to *condemn* me, before you have read what I have to say for myself; and let not the failures of others prejudice me, or be placed to my account. I have often lamented the want of understanding the true powers of Nature, which misfortune has of late put some on making such vast engines and machines both troublesome and expensive, yet of no manner of use; inasmuch as the old engines used many ages past far exceeded them. And, I fear, whoever by the *old causes* of motion pretends to improvements within the last century, does betray his knowledge and judgment; for more than 100 years since, men and horses would raise by engines then made, as much water as they have ever since done—or, I believe, according to the law of Nature, ever will do."[25]

In his address and explanations, Savery proceeds with all the candour and earnestness of a man conscious of having made a discovery of immense importance to mankind; and there is no greater instance of so open and candid an appeal to experiment, and an examination of the actual performance of an engine as a test of its merit, in the history of mechanical inventions: a mode of procedure so opposite to that which could be pursued by a man arrogating to himself the credit due to the genius of another, that we cannot help suspecting Desaguliers of having been influenced by some unworthy personal feeling towards Savery, in his attempt to transfer the invention to the Marquis of Worcester.

Captain Savery, according to the Doctor,[26] "having read the Marquis of Worcester's book, was the first who put in practice the raising of water by fire, which he proposed for the draining of mines. His engine is described in Harris's *Lexicon*; which, being compared with the Marquis of Worcester's description, will easily appear to have been taken from him; though Captain Savery denied it; and, the better to conceal the matter, bought up all the Marquis's books that he could purchase in Paternoster Row and elsewhere, and burned them in the presence of a gentleman who told me this. He said that he found out the power of steam by chance, and had invented the following story to persuade people to believe it; *viz.*: that having drunk a flask of florence at a tavern, and thrown the empty flask upon the fire, he called for a bason of water to wash his hands; and perceiving that the little wine left in the flask had filled up the flask with steam, he took the flask by the neck, and plunged the mouth of it under the surface of the water in the bason; and the water of the bason was immediately driven up into the flask by the pressure of the air. Now, he never made such an experiment then, nor designedly afterwards; which I thus prove:

I made the experiment purposely with about half a glass of wine left in a flask; which I laid upon the fire till it boiled into steam: then pulling out a thick glove to prevent the neck of the flask from burning me, I plunged the mouth of the flask under the water that filled the bason; but the pressure of the atmosphere was so strong, that it beat the flask out of my hand with violence, and threw it up to the ceiling. All this must have happened to Captain Savery: if he ever had made the experiment, he would not have failed to have told such a remarkable incident, which would have embellished his story."[27]

This grievous charge, it has been well observed by Dr Robison, ought to be substantiated by very distinct evidence. "Yet Desaguliers produces none such; and he was too late to know what happened at the time. His argument is a very foolish one, and gives him no title to consider Savery's experiment as a falsehood; for it might have happened precisely as Savery relates, and not as happened to Desaguliers. The fact is, Savery obtained his patent in 1698, *after a hearing of objections*, in which the discovery of the Marquis of Worcester was not mentioned; but, besides this, he had erected several of his engines before he obtained his patent," and published an account of his engine in 1696, under the title of *The Miner's Friend*, and *a Dialogue*, by way of answer to the objections which had been made against it in 1699: both were printed in one volume, in 1702. Here, then, every possible publicity was given, not only to the *principle* of his machine, but, to its construction; and yet during Savery's lifetime the Marquis of Worcester's description had never been mentioned. Neither is this tale of destroying the books found in any other author; nor is it stated by Desaguliers in his own work, which was published in 1734; but in that volume which he first published in 1746, nearly thirty years after Savery's death, and nearly fifty years after the grant of the patent! It is remarkable, too, in the Doctor's account, that the Marquis of Worcester's *book* is mentioned. Was Desaguliers ignorant of the Marquis having published *two* separate books containing the description? If he were, how did the other book become scarce? Copies of that might have been in existence, to be produced in judgment against Savery's claim, and against his patent. It is certain that Savery, during his life-time, had no competitor in England to dispute with him the honour of inventing the machine which now bears his name. But a pamphlet being rare on booksellers' shelves thirty-five years after its publication, is not at all an extraordinary circumstance, and it would, indeed, have been a miracle had a copy, of any equally unimportant book, been found at such a distance of time in that unenviable situation.[28]

Savery exhibited a model of his engine before King William at Hampton Court; and the success of the experiment appeared so satisfactory, that the King warmly interested himself in the project. It was not until June 1699, a year after he had obtained his patent, and erected some engines, that he made trial of his engine before the Royal Society. In his address to that body, prefixed to his *Miner's Friend*, in 1702, he mentions the great "difficulties and expense he had incurred to instruct persons to frame his machine; but that he had at last succeeded in getting workmen, who would oblige themselves to make engines, exactly tight and fit for service; so that he could warrant them to those who chose to employ them."[29]

"The first thing," says the ingenious inventor, "is to fix the engine (Figure 10) in a good double furnace, so contrived that the flame of your fire may circulate round, and encompass your two boilers, as you do coppers for brewing. Before you make any fire, unscrew G and N, being the two small *gauge pipes* and cocks belonging to the two boilers, and at the holes fill L, the great boiler, two thirds full of water, and D, the small boiler, quite full. Then screw in the said pipes again as fast and as tight as possible. Then light the fire at *b*; and when the water in L boils, the handle of the regulator, marked *z*, must be thrust from you as far as it will go: which makes all the steam rising from the water in L pass with irresistible force through *o* into *r*, pushing out all the air before it, through the clack *r*, making a noise as it goes; and when all is gone out, the bottom of the vessel *r* will be very hot.

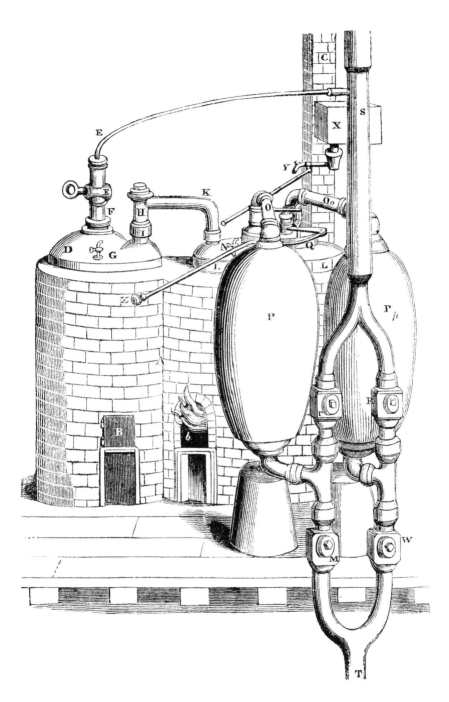

Fig. 10: Savery's engine

Then pull the handle of the regulator to you, by which means you stop o, and force your steam through Oo into Pp, until that vessel has discharged its air through the clack R up the force-pipe s. In the mean time, by the *steam's condensing* in the vessel P, a vacuum or emptiness is created, so that the water must and will necessarily rise up through the sucking-pipe T, lifting up the clack M, and filling the vessel P.

" In the meantime, the vessel Pp being emptied of its air, turn the handle of the regulator from you again, and the force is upon the surface of the water in p; which surface being only heated by the steam, it does not condense it, but the steam gravitates or presses with an elastic quality like air, still increasing its elasticity or spring till it counterpoises, or rather exceeds, the weight of the water ascending in s, the forcing-pipe, out of which the water in it will be immediately discharged when once gotten to the top, which takes up some time to recover that power; which having once got, and being in work, it is easy for anyone that never saw the engine, after half an hour's experience, to keep a constant stream running out the full bore of the pipe. On the outside of the vessel, you may see how the water goes out as well as if the vessel were transparent; for as far as the steam continues within the vessel, so far is the vessel dry without; and so very hot, as scarce to endure the least touch of the hand. But as far as the water is, the said vessel will be cold and wet where any water has fallen on it; which cold and moisture vanishes as fast as the steam in its demerit takes place of the water; but if you force all the water out, the steam, or a small part thereof, going through it, will rattle the clack, so as to give sufficient notice to pull the handle of the regulator to you, which, at the same time, begins to force out the water from Pp, without the least alteration of the stream; only sometimes the stream of water will be somewhat stronger than before, if you pull the handle of the regulator before any considerable quantity of steam be gone up the clack R: but it is much better to let none of the steam go off (for that is but losing so much strength), and is easily prevented by pulling the regulator some little time before the vessel forcing is quite emptied. This being done, immediately turn the cock or pipe Y of the cistern X on P, so that the water proceeding from X through Y (which is never open but when turned on P, or Pp, but when between them is tight and stanch)—I say, the water falling on P, causes, by its coolness, the steam (which had such great force just before, from its elastic power, to condense, and become a vacuum or empty space), so that the vessel P is, by the external air, or what is vulgarly called suction, completely refilled, while Pp is emptying. Which being done, you push the handle of the regulator from you, and throw the force on P, pulling the condensing-pipe over Pp, causing the steam in that vessel to condense, so that it fills while the other empties. The labour of turning these two pairs of the engine, *viz.* the *regulator* and *water-cock*, and tending the fire, being no more than what a boy's strength can perform for a day together, and is as easily learned as their driving of a horse in a tub-gin; yet after all, I would have men, and those too the most apprehensive, employed in working the engine, supposing them more careful than boys.

"In case it should be objected, that the boiler must in some certain time be emptied, so as the work of the engine must stop to replenish the boiler, or endanger the burning out, or melting the bottom of the boiler: to obviate this, when it is thought fit by the person tending the engine to replenish the great boiler, which requires an hour and a half, or two hours time to the sinking of one foot of water, then, I say, by turning the cock E of the small boiler D, you cut off all communication between the great force-pipe S and the small boiler D; by which means D grows immediately hot, by throwing a little fire into B, and the water of which boils, and in a very little time it gains more strength than the great boiler; for the force of the great boiler being perpetually spending and going out, and the other winding up, or increasing, it is not long before the force in D exceeds that in L; so that the water in D, being depressed by its own steam or vapour, must necessarily rise through the pipe H, opening the clack I, and so go through the pipe K into L, running till the surface of the water in D is equal to the bottom of the pipe H. Then, steam and water going together will, by a noise in the clack I, give sufficient assurance that D has discharged and emptied itself into L, to within eight inches of the bottom; and inasmuch as from the top of D, to the bottom of its pipe H, is contained about as much water as will replenish L one foot. Then you open the cock I, and re-fill D immediately, so that here is a constant motion, without fear or danger of disorder or decay. If you would at any time know if the great boiler be more than half exhausted, turn the *small cock* N, whose pipe will deliver water, if the water be above the level of its bottom, which is half way down the boiler; if not, it will deliver steam. So likewise it will shew you if you have more or less than eight inches of water in D, by which means nothing but a stupid and wilful neglect, or mischievous design, carried on for some hours, can any ways hurt the engine. And if a master is suspicious of the design of a servant to do mischief, it is easily discovered by these *gauge pipes*; for if he come when the engine is at work, and find the surface of the water in L, below the bottom of the gauge-pipe N; or the water in D below the bottom of G; *such a servant deserves correction*: though, three hours after that, the working on would not damage or exhaust the boilers. So that, in a word, the clacks being, in all water-works, always found the better the longer they are used; and all the moving parts in our engine being of like nature, the furnace being made of Stourbridge or Windsor brick, or fire-stone; I do not see it possible for the engine to decay in many years; for the clacks, boxes, and mitre-pipes, regulator, and cocks, are all of brass, and the vessels made of the best hammered copper, of sufficient thickness to sustain the force of the working the engine. In short, the engine is so naturally adapted to perform what is required, that even those of the most ordinary and meanest capacity may work it for some years without its receiving any injury, if not hired or employed by some base person on purpose to destroy it." The same letters refer to the same parts in Figures 10 and 11.—Figure 11 represents the apparatus on the top of the boiler L, on an enlarged scale.

This engine Savery applied for raising water for palaces, gentlemen's seats,[30] draining fens, and supplying houses with water in general, and pumping water from ships:[31] and he erected many of them in different parts of England. The power of his engine he limited only by the strength of the pipes and vessels; "for," he says; "I will raise you water 500 or 1000 feet high, could you find us a way to procure strength enough for such an immense weight as a pillar of water that height; but my engine at 60, 70, or 80 feet, raises a full bore of water with much ease." And comparing this performance of his machine with that by manual labour, he continues, I have known, in Cornwall, a work with three lifts, of about eighteen foot each, lift and carry a 3½ inch bore, that cost forty-two shillings a-day (reckoning 24 hours a-day,) for labour, besides the wear and tear of engines, each

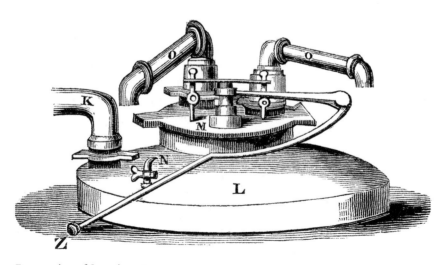

Fig. 11: valves of Savery's engine

pump having four men working eight hours, at fourteen-pence a man, and the men obliged to rest at least a third part of that time. I dare undertake that the engine shall raise you as much water for eight-pence, as will cost you a shilling to raise the like with your old engines in coal pits, which is thirty-three pounds six shillings and eight-pence saved out of every hundred pounds; a brave estate gained in one year out of such great works! where 3, 6, or it may be £8000 per annum is expended for clearing their mines of water only, besides the charge and repair of engines, gins, horses, &c."[32]

Except stating that an engine raising a column of water sixty feet high and three and a half inches in diameter, requires a fire-place twenty inches deep, and fourteen or fifteen inches wide, Savery gives us no information of the proportions of his engines in the *Miner's Friend*. The engraving in that book cannot be depended upon for the correct proportions of the details. The well-known Bradley, who was professor of Botany at Cambridge, gives a description[33] of a small engine having only one receiver, erected by Savery himself about 1711, for a Mr Ball at Campden-house, Kensington. It was standing in Switzer's time, who says it was the best proportioned of any he had seen.

The pipe *a*, in Figure 12, is sixteen feet long, from the surface of the water to the stage on which the receiver *b* is placed. And this is the height to which the water is raised by the pressure of the atmosphere. The height of the reservoir above the receiver is about forty-two feet; and this column of water was elevated by the elasticity of the steam. The pipe *d*, is three inches in diameter, and the steam pipe *e,* about one inch bore. The receiver holds thirteen gallons, and the boiler thirty-nine gallons.

When this engine was in action it raised fifty-two gallons of water in a minute, (four times the contents of the receiver,) and upon an average its effect was greater in proportion than that of the engines with double receivers. "The prime cost of such an engine," says Switzer, "was about fifty pound, as I myself had it from the ingenious author's on mouth, and the quantity of coals required to work it about half a peck, which need not be renewed above six or eight times, were it to be wrought the whole twenty-four hours. Which, supposing to be a bushel at most, is not above twelve pence in London, but much cheaper in many other places. The expense is not considerable to what horse-work is, which must be shifted twice or thrice a-day. The chief thing that seems to be objected against this engine, as to the expense, is the making the fire in the open air as it were and under a trivet. Because the heat in such a latitude will evaporate, and not be so strong as when confined into a narrow compass, and consequently there must be a greater expense of wood and coal than when it is thus contracted,

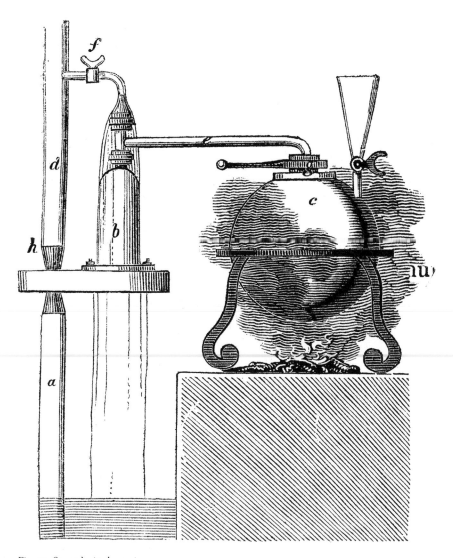

Fig. 12: Savery's single engine

which makes it, I think, better to have the fire enclosed in a stove or furnace thin under any open spherical figure."[34]

Its operation is the same as that of the machine shown in Figure 10,—the steam admitted from the boiler *c,* into the receiver *b,* is condensed by turning the cock *f,* which allows the cold water to fall on the outside of the receiver, at the same time that the flow of steam from the boiler is shut off by the cock *g*; a vacuum being thus produced in the receiver, the pressure of the atmosphere raises the water, in the reservoir up the pipe *a,* and fills the receiver *b.* The cock *f* is now closed, and the communication between the boiler *c* and the receiver *b* is again opened by turning the cock *s*; the elasticity of the steam then forces the water in the receiver *b* up the pipe *d*; the valve at *h,* opening upwards, prevents it from returning. When the receiver is again filled with steam, the cock *g* is shut, and the cold water or condensing cock *f* is opened. This condenses the vapour, and forms a vacuum, and the pressure of the atmosphere again acts to raise the water in the reservoir through the pipe *a* into the receiver, which is forced up the pipe *a* by the elasticity of the steam, and so on alternately.

Some standard of reference was necessary to give definite information of the effect of this new engine; and Savery introduced the term *horse's power*; which is still in very general use. A certain number of horses were kept to raise a certain quantity of water to a certain height; so a steam engine, on Savery's construction, was called a one, or a two, or a three-horse engine, as it raised the water which had hitherto been raised by one, or two, or three horses.[35] We have no means of judging of the data he went on in his calculation of the proportions of the several parts of his machines; but from some circumstances, which we shall afterwards notice, it is extremely probable that they were constructed more from tact than calculation.

While Savery was endeavouring to improve his engines, and introduce them into England, Amontons, a distinguished member of the French Academy of Sciences, was occupied in experimenting on steam and air; and in 1699, he presented an account to that learned body of an engine of his invention, which he called a *Fire Wheel*.[36] An inspection of Figure 13 will convey a clearer idea of Amontons' design, than any verbal description, however minute and detailed. Indeed, although the principle on which it acts is exceedingly simple, the intricate arrangement of its parts almost precludes a satisfactory description with the assistance of one engraving; and its importance is not so great as to induce us to devote more than one to its illustration. This *Fire Engine* consists of a wheel whose diameter is divided into four concentric circular compartments, or rings. The outermost circle is divided on its circumference into twelve chambers, A, B, C, D, &c. all of which are so formed as to be air-tight, and to have no communication with each other. The second portion of the diameter is open, and serves the purpose of insulating the outer series of chambers, from a corresponding series formed in the third portion of the wheel, and marked a, b, c, d, e, f, &c. These chambers communicate with each other, by means of valves or flaps, which open only in one direction (upwards.) The inner portion of the diameter is filled up with a series of pipes, and the axis or gudgeon of the wheel z, on which it revolves. W is a cistern of cold water, in which the under portion of the wheel is immersed. The fire-place, and the mode in which the flame acts on the outer surface of the first series of chambers, will he clearly understood by a reference to Figure 13. X is the chimney, by which the smoke and heated vapour escapes, after it has come into contact with the periphery of the wheel.

The outer series of chambers, A, B, C, &c. is connected with the inner series marked a, b, c, d, &c. by means of the pipes 1, 2, 3, 4, &c. The pipe, 1, connects the air-chamber, A, with the inner water-chamber, a. The pipe, 2, connects the air-chamber, B, with the inner water-chamber, b, and so on round the series. These pipes marked, 1, 2, 3, &c. are inserted into another pipe closed at one end and open at the other; the open end of the enclosing tube is inserted into the water-chamber, forming the communication between the inner and outer series of chambers.

If we now suppose the cells, a, b, m, in the second row to be nearly filled with water, and the outer side of cell A, heated by being placed over the fire, the air which it contains being expanded by the heat, issues through the pipe 1, into the cell a, of the second row of chambers, which is filled with water. It presses on the surface of this water, and raises it into the cells b and c, (being prevented from passing into m, by the valves or flaps opening from the right to the left hand in our engraving,) and into d and e, in proportion to the degree of rarefaction of the air. The weight of the water, raised into b, c, and d, gives that side of the wheel a preponderance. The air-chamber A, is consequently depressed into the position of m, and the cell h, is then exposed to the action of the fire. This air-cell communicates with the water-chamber, b, by the pipe 2; cell b, being placed in the position of a. The air in chamber B being expanded by the heat, issues through pipe 2, and presses the water in b, into e, d, &c.: which continues the preponderance. The air-chamber C, is next exposed to the fire, and so on through the series continually. When the surface of the cell c, has reached the furnace, the cell A, is then (in the position of L in the figure) immersed in the cold water of the cistern w, which cools and condenses it. The heated air in cells B and C, is also condensed as

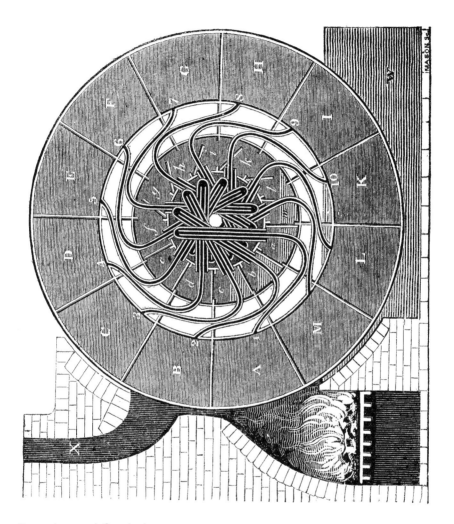

Fig. 13: Amonton's fire wheel

they move round in the water; and emerging, they are again placed in succession in contact with the fire. This rarefaction of the air, forcing water up on one side of the wheel, gives it a successive preponderance which produces a rotatory motion, that may be applied to practical purposes.

Amontons gave a diameter of twelve feet to the inner row, or water-cells, and he made them two feet deep, or wide, enclosing about 754 cubic feet of water, about 48,202 pounds (French) weight. This applied tangentially to a circle supposed to pass through the centre of the water-cells, would, according to his calculation, make a revolution in thirty-five seconds, and be equal in power to thirty-nine horses.

The *Fire Wheel* though extremely ingenious, is by far too complicated for any practical purpose; even granting that it were possible to produce the degree of rarefaction in the air-chambers to raise the column of water, with the assigned velocity; and that the action of the valves was practicable or could be depended on. The inventor describes its action as produced solely by the rarefaction of the air; but its power would in practice be derived from the expansion of *steam*: for no ingenuity could in this arrangement prevent the vapour, formed in the water-cells by the heated air, from passing into the outer series of air chambers by the pipes. And this circumstance might give to a similar apparatus that power, which it would clearly not possess by the expansion of air alone, as proposed by the inventor.

In August 1705, M. Dalesme, the well-known author of the ingenious mode of consuming smoke by reversing the flame, offered to a company in Paris, to raise water by means of steam issuing from, an engine similar to an *oelipile*. In his experiments he applied the elasticity of the vapour to make the water to spout to a great height. Prony[37], in regretting his ignorance of the details of Dalesme's method, thinks "his model may still be extant in the collection of the machines of the Academy; but this cannot be ascertained until that collection is put in order." This was written thirty years ago.

Dr Papin, whose ingenious project for forming a vacuum under the piston of an air-pump we have already noticed, was still employed in (1698) in making experiments on steam, under the patronage and at the expense of the Elector of Hesse. In discontinuing those researches, which had not led him to any useful combination, Papin communicated the result, with some suggestions, to several of his literary correspondents,—among others, to the celebrated Leibnitz. In his reply to Papin's letter, Leibnitz mentioned that the idea of employing the expansive force of steam had also occurred to himself; and during a subsequent visit to England, in 1705, having had an opportunity of inspecting some of the machines which were erected by

Savery, he sent a sketch with a description of one of them to Papin, to have his opinion on the merit of the invention. This letter and sketch were shown by Papin to his patron the Elector. At the command of that prince, Papin, to use his own terms, "resumed his experiments, to complete those contrivances for raising water by fire which he had commenced so favourably." The result of this application was a publication by the Doctor of an account of a "New method of raising water by the force of fire," dated at Cassel in 1707, and dedicated to the Elector. In this production, Papin states the invention to have been perfected after a *great many failures*; and with a refinement in flattery, he declares that the printing of the account of this machine was not done to detract from the merit of Savery's engine, (for he acknowledges Savery's having hit on another mode, *without knowing of his experiments*;) but that the whole world should know that it was indebted to the Elector of Hesse for the first idea and construction of this excellent contrivance. But it is worthy of remark that in the body of the treatise Papin's endeavours appear to be directed much more to showing the superiority of his apparatus over that of his rival, than to establish a claim to priority of invention. "His forty pages of calculation prove his understanding and learning; they have, however, failed in their object of convincing practical men to adopt this engine."[38]

A boiler, a, made of copper, communicates by a pipe z, with a cylinder i, which forms the body of the pump. This cylinder has no bottom, but is attached by a curved pipe x, to an upright pipe o, q, which enters the cylinder, $r\ r$, rising to within a short distance of its top. This cylinder is air-tight, and has a pipe w, and stopcock, p. On the curved pipe x is fixed another pipe, m, terminating in a funnel or reservoir, k, with a stopcock at m. The pipe z, joining the boiler and pump cylinder, has a stopcock at c, and another small pipe at e, also furnished with a stopcock; f is a valve, introduced by Papin, which has incalculably increased the security from accident of boilers for the purpose of generating steam. The fire is made at b; within the cylinder i is a piston or float n, made of thin plates of metal, forming a part of a hollow cylinder, which floats on the surface of the water: d is a pipe and stopcock inserted into the vessel i.

When a sufficient quantity of steam is generated in the boiler a, the cock c is opened, to allow it to flow into the pump-cylinder i, which we may suppose to be nearly filled with water. The steam presses by its elasticity on the floater n, and depresses it in the cylinder i, forcing the water, which is beneath it, through the curved pipe x, up the perpendicular pipe o, q, until it falls at the upper end q into the receiver r, r; this condenses the air in r, r. The stopcock p, is then opened, and the water, pressed by the condensed air in the upper

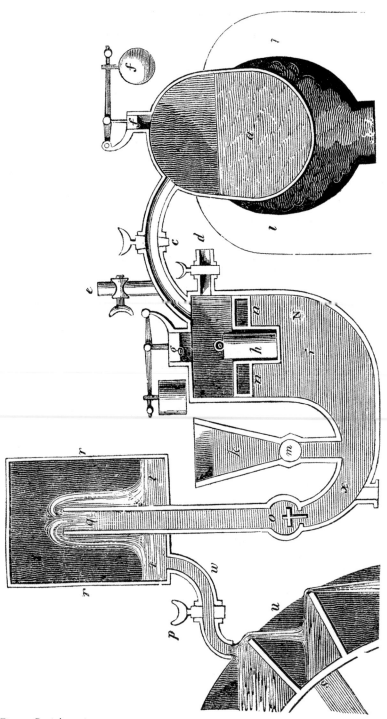

Fig. 14: Papin's engine

part of the cylinder, issues with great velocity upon the float-boards *u, s, x*, of a wheel, which is carried round and imparts its motion by the usual intermediate means to pumps or other machinery. When the floater *n* has reached the situation N, the cock *c* is turned, which prevents the further admission of steam from the boiler into the pump- cylinder above the floater; and the cock *d* is opened to allow the steam above the floater to escape into the atmosphere; at the same moment the stopcock *m* is opened, which admits the water in the funnel, *k*, to descend through the curved pipe *v*, and to raise the floater *h,* to the necessary height, in the pump-cylinder *i*; the clack at *o*, opening upwards, prevents the column of water in *o, q,* from descending. When the floater has risen to the proper position, the cocks *d* and *m*, are shut, and *c* is opened. The steam again flows from the boiler into the cylinder, and depresses the floater *h*; raising the water under it into the reservoir *r, r*; and when it is depressed to the situation N, the cock *c* is shut, and *m* and *d* are again opened, to perform the operations of filling the cylinder with water, elevating the piston or floater, and extruding the steam into the atmosphere, and so on continually, which keeps a constant stream falling from the pipe *w*, with a considerable velocity on the float-boards of the water-wheel, *u, s, x*. The pipe and cock *e*, are for allowing the air in the boiler to escape when it is first filled with steam; *n*, is a similar pipe and cock for emptying the cylinder and pipe *x* of water.

It does not, however, appear that Papin adopted the floater to prevent the condensation of the steam, by keeping it from coming into contact with water; for it was practically impossible to produce this contact in his apparatus. His purpose, very different, has been totally overlooked by all English writers who have described his contrivance. One of Papin's projects was to increase the "force of the steam" by introducing a mass of red-hot iron into the pump-cylinder. This beater was placed in a pipe fitted into the floater, closed at the bottom to prevent the access of the water. *n, n,* is a section of the floater; *h,* is the heater suspended in the pipe, having a ring to withdraw it by when it became cold; and *g*, is the opening in the top of the cylinder by which the heater is introduced, or taken out. The opening is closed with a lever valve, retained in its situation by the weight hung at its extremity. This heating contrivance presents, according to Prony,[39] so many difficulties, that it is doubtful if it could be put into practice; an opinion entirely coinciding with our own, yet without going the length to condemn the *whole machine* (with Dr Robison), as "so awkward, and so unlike any distinct notions on the subject, that it would not do credit to any person;"—an assertion very much at variance with the fact. The mode of elevating the piston, and of producing a constant steam of water through the pipe *p*, are both distinct and ingenious notions. The introduction of the *Safety Valve* may even now be considered as one of the most important improvements made in Steam apparatus; and the want of which long and powerfully retarded the introduction of Savery's engine.

"Bossut,"[40] continues the Professor, "says that the first notion of the Steam Engine was *certainly* owing to Dr Papin, who had not only invented the digester, but had in 1695 published a little performance describing a machine for raising water, in which the pistons are moved by the vapour of boiling water alternately dilated and condensed. Now the fact is, that Papin's *first* publication was in 1707, and his piston is nothing more than a floater on the surface of the water, to prevent the waste of steam by condensation; and the return of the piston is not produced, as in the steam engine, by the condensation of the steam, but by admitting the air and a column of water to press it back into its place." We scarcely know what to make of a statement which is so utterly irreconcilable with fact. Bossut evidently refers to Papin's project of employing *steam* as a substitute for gunpowder, an account of which *was published* by its inventor, *a second time*, in 1695. But Dr Robison says Bossut *means* the engine for raising water by the *elasticity* of *steam,* of which there was *no* description printed before 1707, eleven years afterwards!

The Professor's opinion, that Papin's employment of a floater in his engine described in 1707, (Figure 14) gives him no claim to the first invention of a piston, is so far correct; but it does not follow that a previous claim for a totally different invention is also groundless; and it so happens that it is on account of an earlier suggestion, which is also truly stated by the Doctor himself, that Bossut, and we believe every one else, gives Papin the merit of having suggested the *Atmospheric Engine*. In describing the machine which bears his name, the most eminent French engineers have done ample justice to Savery, as its inventor.[41]

Indeed, throughout the discussion of Papin's invention, Dr Robison's usual candour and liberality seem to have forsaken him; and in noticing an engine, allowed even by its projector to have failed in practice, he passes a crude and hasty opinion on an improved construction of the same machine, which, for its ingenuity and importance, may rank next to the Steam Engine itself. It is true that in the Professor's time, Papin's mode of transmitting power and motion to great distances was not applied to any useful purpose; but it has recently been introduced with admirable effect into a mechanism employed for an important national purpose. This is, however, among those stated by Dr Robison to be "awkward, absurd, and impracticable inventions. Papin's conceptions of natural operations were always vague and imperfect, and he was neither philosopher nor mechanician."[42] Practical men are more just in their opinion of Papin; they place this ingenious Frenchman in the highest rank of scientific mechanics.[43]

The advantage derived from the use of Savery's engine as a substitute for manual labour was counterbalanced, in public opinion, by the great risk of accident from an explosion of the boiler: for, during the term of his patent, it does not appear that he availed himself of the security arising from the use of Papin's safety-valve. At York Buildings' water works, Desaguliers[44] knew Savery to make steam eight or ten times stronger than the common air; and then its heat was so great that it would melt common soft solder; and its strength so great as to blow open several of the joints of his machines, so that he was forced to be at the pains and charge to have all his joints soldered with spelter or soft solder. Various attempts were made to strengthen the boilers, by radiating arms fixed in the inside; but without any successful result: so that at this period the only use to which Savery's apparatus could be applied with safety, was to raise water to heights not exceeding 30 or 32 feet—a virtual abandonment of its pretensions as a mine-draining power, which was the grand object of all Savery's exertions.

Notwithstanding this failure, the introduction of these machines in the mining districts had a most important influence in directing attention to the useful properties of elastic vapour. The variety of interests and opinions opposed to, or in favour of their use, exciting discussion, rendered the mining population familiar with the more obvious laws of the elasticity of steam, and the effects of its condensation.

Among those whose attention was more immediately drawn to the vast importance of Savery's machine, as developing a power alone capable of preserving a vast property from ruinous deterioration through the accumulating difficulties of an extending and necessary drainage, were a Thomas Newcomen, a blacksmith, and John Cawley, a glazier, both living in the town of Dartmouth in Devonshire. At this distance of time it is impossible to ascertain what share Cawley had in the invention or experiments; but from some papers of Dr Hooke[45], it appears that Newcomen had been in correspondence with the Doctor, respecting a project to produce a moving power on Papin's plan of an *air pump*. Among Hooke's papers there are some memoranda of a letter addressed to Newcomen, dissuading him from erecting a machine on that principle, in which there is the following remarkable expression: "Could he (Papin) make a *speedy vacuum* under your second cylinder, your work is done." Dr Hooke knew of Papin's scheme of making a vacuum by steam under the piston, and in some discourses before the Royal Society he demonstrated its impracticability; and it can scarcely be imagined to have been unknown to Newcomen. However, the effect of the condensation of steam and its elastic power were now well known; and it is probable that the success of Savery's experiments opened new views to Cawley and Newcomen, which decided them in favour of the practicability of Papin's project. They, therefore, made the experiment of introducing *steam* under *a piston moving in a cylinder, and formed a vacuum by condensing the steam by an affusion of cold water, on the outside of the steam vessel; and the weight of the atmosphere pressed the piston to the bottom of the cylinder.* This was the first form of the Atmospheric Engine,—the simplest, and the most powerful machine that had hitherto been constructed.

The method of procuring a vacuum by the condensation of steam was, however, employed in Savery's engine, and secured to him by patent.[46] An arrangement was therefore entered into between Savery, Newcomen, and Cawley, and all their names were associated in the grant of a monopoly for the new engine which was made in 1705. "As the best and most useful inventions and improvements," says Switzer,[47] "which have been discovered either in art or nature, have in the process of time been subject to the same, so this ingenious gentleman (Mr Newcomen), to whom we owe this late invention, has with a great deal of modesty, but as much judgment, given the finishing stroke to it. It is, indeed, generally said to be an improvement to Mr Savery's engine; but I am well informed that Mr Newcomen was as early in his invention as Mr Savery was in his, only the latter, being nearer the court, had obtained his patent before the other knew it; on which account Mr Newcomen was glad to come in as a partner to it."[48]

In Savery's Engine we have seen that the effect is produced in two ways, by the condensation of the steam, forming a vacuum in a receiver, into which the water is forced by the pressure of the atmosphere; and where the water was required to be elevated to a greater height than from 28 to 30 feet, he employed the direct pressure of steam of a high temperature, and dangerous elasticity.

In the Atmospheric Engine, the process is totally different; the steam exerts no *direct* action upon the water, or upon any part of the apparatus,— it is merely employed as a means of forming a *"speedy vacuum,"* under a piston attached to one end of a lever, the rod of a pump-piston, or plunger, being affixed *to the other extremity.* In this construction the power of the engine has no reference whatever to the *strength* or *temperature* of the steam, but depends upon the superficial dimension of the piston beneath which the vapour is introduced from the boiler. The steam cylinder, T, was now, for the first time, effectually detached from the water-pump.

The steam generated in a boiler, *b*, in Figure 15, was admitted through the cock *d*, and pipe *q,* into a cylinder *a,* under the steam piston *s,* attached by the rod *r*, to the lever or beam *i, i,* moving on the axis or fulcrum *o*. The cylinder *a* was placed in another cylinder, forming a concentric space, *z, z,* round it. This outer cylinder was connected by a pipe, *f* to a reservoir, *g,* containing cold water. Another pipe proceeding from its lower end was inserted into the well or second reservoir of cold water.

The piston being in the position shown in the Figure, and the cylinder *a,* being filled with steam through the pipe *q,* the cock *d,* is turned, which shuts off the communication between the cylinder *a,* and boiler *b*. By opening cock *f,* cold water is now allowed to flow from the reservoir g, through the pipe *f,* into the outer cylinder *z, z*: this cools the cylinder *a,* containing the steam, which condenses the included vapour, and forms a *vacuum* under the piston *s*. The pressure of the atmosphere meeting with no resistance from the elasticity of the steam, forces the piston to the bottom of the cylinder.

By this movement, the end of the lever *i*, attached by the rod *r*, to the piston, is depressed; and the other end of the lever to which the pump-rod is fixed, is raised, and draws up all the water above the plunger in the pump barrel along with it.

Now, if we suppose the cold water which has been in contact with the steam cylinder to have condensed all the vapour, the atmosphere will press on the piston with a force equal to that which would be produced by placing about 14¾ lbs weight on each inch of its surface. If the piston were 62 inches square, this would be about 915 pounds weight, operating to force it downwards; and, if there were no resistance from friction, it follows, that in the same time an equal weight placed at the other end of the lever beam, or a column of water, weighing 915 pounds, would be lifted as high as the steam piston had been depressed in the cylinder.

When the piston has arrived at the bottom of the cylinder, the cock *d* is turned, which again opens a communication between the boiler *b*, and the steam cylinder *a*. In this engine, the steam being only equal to the pressure of the atmosphere, the piston, *s*, must be raised by other means to the top of its cylinder. This is effected by a counterpoise, *m*, fixed on *k*, which is so adjusted as to depress the pump rods, and thus to raise the steam piston into the required position. During this operation the cock *f* is shut, and the cock *e* is opened; and the water heated by the condensation of the steam, in the condensing cylinder *z*, escapes into the well or tank *o*. A very small quantity of water being formed in the steam cylinder *a*, by the condensation of the vapour, is allowed to fall through the pipe *p*, into the same receptacle. The

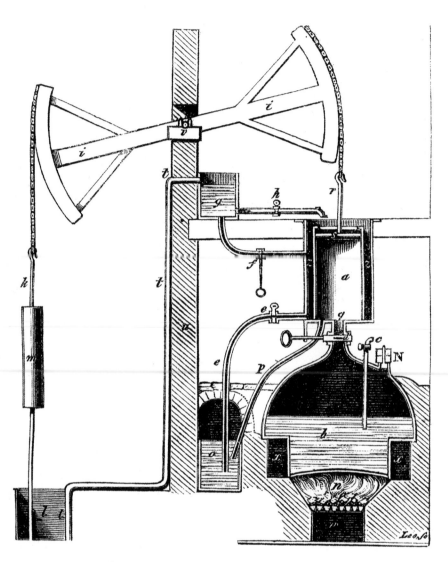

Fig. 15: Newcomen's first engine

cylinder being a second time filled with steam, the cock *f* is opened, and cold water flows from the reservoir *g* into *z*; the steam under the piston is again condensed. The pressure of the atmosphere a second time having the preponderance, the piston is depressed, and the pump rod at the opposite end of the lever beam is elevated, lifting up the column of water in the pump barrel as before. By closing cock *f,* opening *e* and *d,* the counterpoise, *m,* again acts to raise the piston *s*; and the operation may thus be indefinitely repeated.

The fire-place under the boiler is shown at *n,* the ash-pit at *w; x, x,* are the smoke flues; N is a safety valve; c, a *guage pipe,* as in Savery's Engine; *u,* a wall or post supporting the axis *v,* of the lever beam *i,i*; *t, t,* a pipe connected with the pump barrel, in which the cold water rises to supply the reservoir *g; v, 1,* is the mouth of the well or mine which is to be drained; *h,* a pipe proceeding from *g,* through which water flows on the top of the piston to keep it airtight—a contrivance first used by Newcomen.

In this state of the apparatus, at the latter end of the year 1711, the patentees made proposals to draw water from a mine at Griff in Warwickshire; but their scheme did not meet with any encouragement. In March 1712, through the acquaintance of a Mr Potter, of Bromsgrove in Worcestershire, Newcomen succeeded in getting a contract to draw water for a Mr Back, of Wolverhampton. After a great many laborious attempts they made the engine work; but not being philosophers enough to understand the reasons, or mathematicians enough to calculate the powers and proportions of the parts, they very luckily found by accident what they sought for. They were at a loss for the pumps; but being so near Birmingham, and having the assistance of so many admirable and ingenious workmen, they soon came to the method of making the pump valves, clacks, and buckets; whereas they had but an imperfect notion of them before. "One thing was very remarkable:—as they at first were working, they were surprised to see the engine go several strokes and very quick together; when, after a search, they found *a hole in the piston which let the cold water in, to condense the steam in the inside of the cylinder; whereas before they had always done it on the outside.*"[49]

This fortunate observation gave rise to the fine improvement of condensing by injection, which henceforward rendered the outer or water cylinder useless. The pipe, *f,* in Figure 16, proceeding from the cold water reservoir *g,* is inserted into the bottom of the steam cylinder. When the piston, *s,* is at the top of the cylinder, and that vessel filled with steam, the cock, *f,* is opened, and the cold water issues in a jet, and condenses the vapour; a vacuum is formed with prodigious rapidity beneath the piston, and the pressure of the atmosphere gives it a downward motion. The *injection water* escapes by the

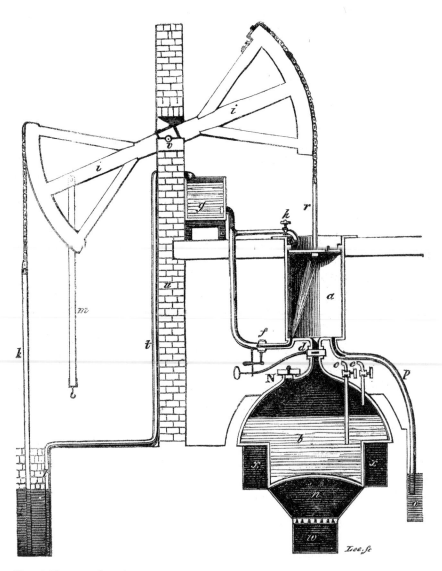

Fig. 16: Newcomen's engine

pipe, *p*, into the well, *o*. The action of all the parts shown in this Figure is the same as those of the last one; and the same letters in each refer to the same parts. *m* is a counterpoise attached to the lever beam, instead of to the pump-rod. The varying lengths of these rods require different weights to be hung at its lower end, for the purpose of adjustment. When the machines were working under loads inferior to their whole power, to prevent shocks arising from this cause, the quantity of injection was lessened, or the injection cock was shut sooner.

They used, continues Desaguliers, to work with a "buoy in the cylinder, enclosed in a pipe; which buoy rose when the steam was strong, and opened the injection and made a stroke: thereby they were only able from this imperfect mechanism to make six, or eight, or ten strokes in a minute; till a boy named Humphry Potter, who attended the engine, added what he called a *scoggan*,[50]—a catch, that the beam (or lever) always opened; and then it would go fifteen or sixteen strokes in a minute."[51]

In consequence of the comparative precision which this improvement gave to its movements, as well as on account of the greater safety, and immense saving which was made in the quantity of fuel, when compared with that used by machines on Savery's principle, the use of the Atmospheric Engine had so extended as in a great degree to supersede the high-pressure engines.

Still many inconveniences remained to be removed; and not the least was the necessity of employing boys or men to open and shut some of the cocks: for, although the risk of accident from the explosion of the boiler might now be considered to have been obviated, the effect of the engine depending much on the condition of its parts, and these being easily deranged by slight irregularities in their action, the danger of injury to the machine itself was considerably increased, from the ignorance or carelessness of the attendants.

The mechanism for opening and shutting the cocks also remained perplexed by catches and strings, until Mr Henry Beighton, an engineer extensively employed in the construction of mining machinery, erected an engine at Newcastle-on-Tyne, in 1718, in which all these "cock-boys" and complication of cords were superseded by a rod suspended from the beam, which operated on a mechanism invented by him called *hand-gear*; a contrivance, with some slight modifications, employed in engines of the present day. It would also appear that the steelyard safety-valve was first used in the boiler of this engine, having been suggested to Beighton by Desaguliers.

The cylinder of the Griff Engine was 22 inches in diameter; and Beighton calculated that it contained 113 gallons of steam at every stroke, equal to about 14,464 gallons per minute, which was produced from about five pints

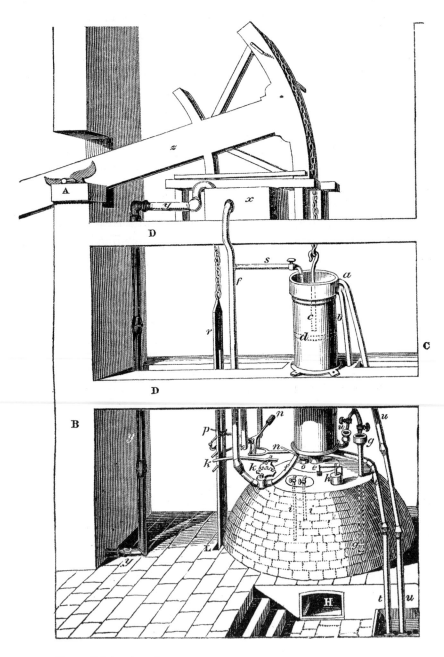

Fig. 17: Beighton's engine

of water; and this quantity was equal in its performance to three-fourths of the atmospheric pressure; so that, making allowance for the friction[52] of the piston, levers, and other parts, about eight pounds of water was raised by each square inch of the piston.

In examining the Figure 17, which is a view of the Atmospheric Engine as improved by Beighton, it will be seen, that, in addition to the hand-gear, he gave a better arrangement and form to the parts already in use, and paid more attention to the proportion of the parts among themselves, and to the work which they had to perform; besides introducing greater neatness and accuracy of workmanship into his engines than had been attempted by his predecessors.

In Figure 17 the cistern, *x*, for the supply of injection water, is placed as in the previous engravings, and water is pumped into it by a small pump connected with the pipe, *y, y*, leading from the mine. (The lever beam, *z*, is not continued on the pump side beyond its axis, A, as this would have required our Figure to have been drawn on a scale much too small for being distinct.) To make the piston, *d*, air-tight, a ring or piece of match[53] is laid upon its circumference; which is kept moist by a small stream of water kept constantly running from the pipe, *s*, upon the piston, *d*;—a projecting rim rising above the highest point to which the piston is elevated, prevents the water from flowing over the sides of the cylinder, when the piston has reached its upward stroke: this will also be observed in the preceding figures. The boiler which is shown as cased in brick-work, is supplied with warm water from this rim by a pipe, *b*; the water falls into a funnel, *g*, attached to a pipe, *g*, which rises to a convenient height above the top of the brick-work, and descends about a foot into the water in the boiler; the two *gauge-pipes, i, i*, are used (as in Savery's Engine,) to ascertain the quantity of the water in the boiler; the lower end of one is immersed for a short distance in the water—the lower end of the other reaching to within a few inches of its surface. If steam issues from both cocks when they are opened, there is a deficiency of water in the boiler; if both give water, then it shows there is an overabundant quantity. The cold water is injected into the cylinder through a pipe, *f*; and after it has performed its office of condensation it is conducted by the pipe, *t, t*, and escapes through a valve at its extremity into the well or reservoir. When the water which flows from *s*, on the top of the piston, is not all used to supply the waste of evaporation in the boiler, its accumulation would soon fill the rim or cup above the piston and flow over its edge upon the casing of the boiler. To prevent this overflow, a pipe, *u, u*, is inserted at *a*, which allows the accumulated quantity to fall into the

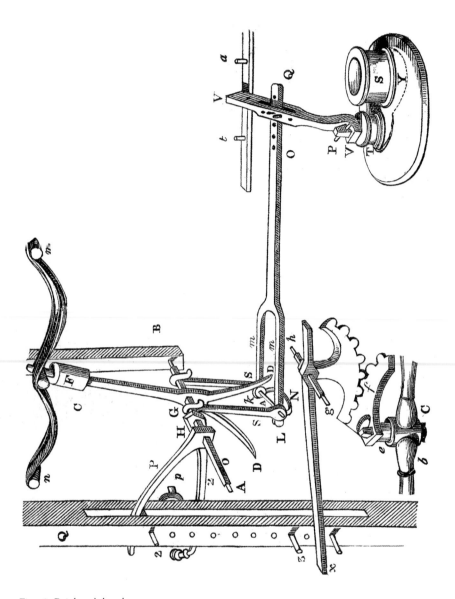

Fig. 18: Beighton's hand-gear

well. The air which is contained in the injected water, and produced by the condensation of the steam under the piston, escapes by a small pipe, *w,* to which is attached a little cup, with a valve opening outwards: when the air is expelled by the descent of the piston, it is shut by the pressure of the atmosphere; a small quantity of water is occasionally introduced by the pipe *n,* to keep it air-tight. "This is called the *snifting-valve,* because the air makes a noise every time it blows through it, like a man *snifting* with a cold."

With the exception of the position of this valve, which Newcomen supplied by the pipe used for conveying the water produced by the condensation of the steam into the well, all the parts that we have particularized have the same operation with those similarly placed in the previous figures. The hand-gear, contrived by Beighton, is shown in the Eighteenth Figure, on a larger scale than in the preceding engraving, for the purpose of giving a clearer view of its construction and action. The Atmospheric Engine, by its introduction, first properly became a self-acting apparatus.

Between two perpendicular pieces of wood, (not to confuse the Figure, one only, B, is shown,) there is a square iron axle, *o,* which has upon it four iron pieces subservient to the turning of the regulator, by shooting forward and drawing back the fork, *m, m,* fastened to the handle, *v, v,* of the *regulator,* T. In the perpendicular working beam, called by Beighton the *plug-frame,* there is a slit which is contrived *o* that its pins work on the fore part, middle, and back part, to raise and depress the levers, *x, p, z,* that move the iron axle, *o,* as much round as is necessary. On the iron axle is fixed a piece called the Y, from its resemblance to that letter, with a moveable weight, *r,* fixed on its upper end. The *stirrup*, *x,* is fixed to the hooks, *s, s,* suspended on the iron axle; the levers or *spanners* are also fixed upon this axle, at right angles to the Y piece. The handle of the *horizontal fork* has holes near its extremity, for the purpose of keeping any part of the end, *a,* in any part of the regulator lever, *v, v,* which moves on a horizontal bar between the pins, *t, a*.

From the situation of the apparatus, the regulator is partly open, which is apparent from the shifting plate or valve, shown by the dotted line *y,* being turned from under the throat-pipe, *s,* which communicates with the cylinder the situation of the piston in the cylinder will be somewhat higher than shown in Figure 17, consequently the lever-beam and the *plug-frame* are nearly at their greatest elevation; and the pin or pulley, 2, in the slit of the *plug-frame,* has so raised the lever or *spanner, p,* that the weight of the head of the Y piece, is brought so far from under *n,* as to have past the perpendicular to the axle, and being ready to fall over towards in, its shank, D, will strike the pin, 4, of the stirrup, with a smart blow, and drawing the fork, *m, m*, horizontally towards the plug-frame, will also draw the end, *o*, of the handle of the regulator, *v*, (which slides on the bar between *t*, and *a*,) and thereby shut off the communication between the cylinder and the boiler. The fall of the plug-frame will reverse this motion. The moment this movement is completed, the pin, 3, on the outside of the plug-frame, depresses the lever, *x*, attached to the quadrant of a wheel, *g*, which moves another quadrant, *f*, which is fixed on the axis or spindle of the cock, *e*, of the injection-pipe, *b, c*. This, admitting cold water into the cylinder, condenses the vapour and produces a vacuum; and the pressure of the atmosphere carries the steam-piston downwards and raises the plug-frame. The lever, *x*, is raised by another pin which shuts the injection-cock, and depressing one of the *spanners* fixed on the iron axle, moves the stirrup and fork into the position which opens the sliding valve, and permits the steam again to issue from the boiler into the cylinder.

Figure 19 is a geometrical view of the same engine, slightly varied in some of its details, and which on the whole may be considered as improvements. As its action is the same as those we have already described, an enumeration of the names of the parts will be sufficient to explain their uses. H, is the fireplace under the boiler, *w; i, i*, are the two gauge-cocks; *o*, the spindle of the regulator valve, which opens or shuts the communication between the cylinder and boiler by the *throat-pipe, e*; the pipe, *t, t*, carries the heated injection water into the well; from this pipe, a small branch, *g*, proceeds with a funnel-cup, having a valve opening upwards; the hot injection water passes from this into the boiler, and an additional supply is procured by the pipe, *b*, from the cup containing the water used to make the steam-piston air-tight; *r, r*, is the *plug-frame*; *p, p*, the *spanners*, moving the fork, and lever of the regulating valve, *o*, which is constructed somewhat differently in this from the preceding Figure; *m*, is the *tumbling bob*, which has the same use and operation as the Y piece. The *injection-cock, k*, is moved by a similar contrivance of a fork acting on the end of a lever, and which is put in action by pins fixed in the *plug-frame*, to move the *spanner, k; e*, is a weight or *tumbling bob*, or Y piece, to give the necessary momentum to the movement of the injection-cock lever; *s*, a pipe from the cold water cistern-pipe, *f*, from which a small stream constantly flows on the top of the piston; *x*, the cold water cistern; *c*, the rod attaching the steam-piston to the chain fixed to the lever beam; *z*, the lever beam; A, its axis; B, the wall or post which supports it; D, beams or joists to which the flinches of the cylinder are bolted; L, the groove or cavity in the floor of the engine-house, in which the lower extremity of the plug-frame moves; F, a cock for emptying the boiler; *h*, the safety-valve.

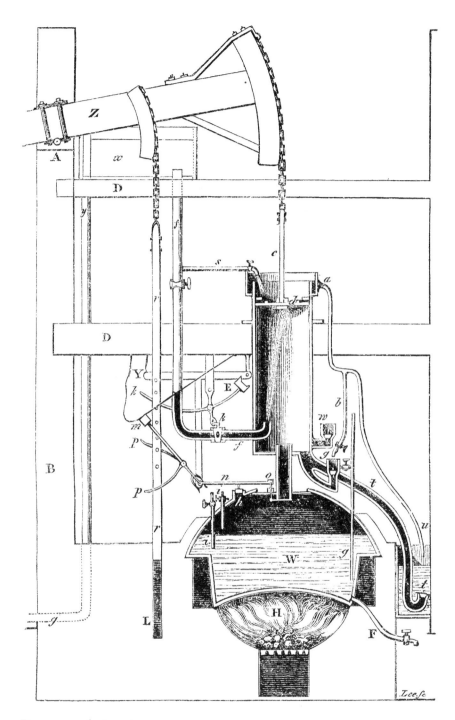

Fig. 19: atmospheric engine

In all the figures we have given of the Atmospheric Engine, the steam cylinder is shown standing upon or over the boiler: this was the usual construction. It was obviously difficult and impossible to keep the boiler and cylinder so firm in their position, as was necessary to preserve the accuracy of their movements. The least want of precision, too, in the opening and shutting of the injection and steam-valves, was attended with the most disastrous effects in this arrangement Engines on the best construction, and most accurate workmanship, were subject to a jolt or shake, at the moment when the vast mass of matter in the beam had its motion reversed, which loosened the boiler in a very few days from the brick-work; and even in the engines constructed by Smeaton, forty years after this period, where every precaution was taken to insure stability which could be suggested by a sagacious mind and long experience,—of bolting the cylinder to strong beams of timber, resting on massy walls totally unconnected with the boiler or its appendages, as in Figures 16, 17 and 19,—yet, during the upward motion of the piston, the boiler and cylinder were lifted up as much as the resilience of the supporting beams would allow, and equally depressed at the downward movement. To obviate these disadvantages, the parts were sometimes arranged differently,—the cylinder was placed *beside* the boiler, and it was fastened to a solid mass of wood and stone, quite unconnected with that vessel. This was a decided improvement; but still it was found that the pipes connecting the boiler and cylinder were almost always out of order, and were a constant source of repair and its attendant expenses.

"In 1716," says Desaguliers, "when Dr S. Gravesande (who had come to England as secretary to the Dutch Embassy) and myself were considering Savery's Fire Engine, as it is described in Harris's *Lexicon Technicum*, it appeared to us that there was a great waste of steam, by its continually acting upon the receivers without intermission; it becoming useless until it had heated the surface of the water in the reservoir, and also to a certain depth: we thought, that were it so contrived that, after the steam had pressed up one receiver full of water, instead of being thrown upon another, it should be confined in the boiler till the reservoir was refilled by the atmosphere, and then turned upon the water; that by this means its confinement might give it so much force, that it would push hard against the surface of the water, and have discharged a great deal of it even before it had heated the surface; and that Savery had, perhaps, in his great work chosen to use two receivers, because the Marquis of Worcester mentions two in his account. We resolved to have a working model, made to act either with one or two receivers. This model soon shewed us that one receiver could be emptied

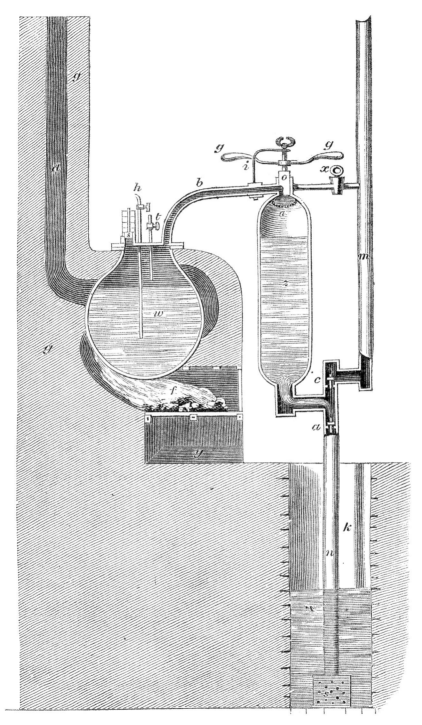

Fig. 20: Desaguliers' engine

three times, whilst two succeeding ones could be emptied but once each; so that, by this means, an engine would be so simple as to be more easily worked, cost almost half less, and raise a third more water."[54]

This engine is shown in Figure 19. A spherical boiler, *w*, is enclosed in brick-work; *f*, is the fire-place, and *g*, the ash-pit; *d*, the chimney; *t, t*, are the usual gauge-cocks; *r*; is the water in the reservoir *k*, which is to be raised through the suction-pipe, *n*, and valve *a*, into the receiver *z*; the pipe, *b*, connects the boiler and receiver; *o*, a cock called a four-way cock, with levers or handles, *g, g*, for opening it; *x*, a cock and pipe, conveying water for the injection from the eduction-pipe, *m*. The operation is the same as of that shown in the Twelfth Figure. The steam flowing from the boiler through *b*, is admitted by the cock, *o*, into the receiver, *z*, and pressing on the surface of the water, raises it through *c*, up the pipe *m*. The lever or handle, *g*, is now turned, and the cock presents its solid side to the opening of the pipe *b*, which interrupts the communication between the boiler and receiver, at the same time that a quantity of cold water is allowed to fall into the receiver from the pipe, *x*; this condenses the steam, and forms a vacuum in the receiver; the water in the reservoir, *k*, is now elevated by the atmospheric pressure through pipe, *u*, into the receiver, *z*, and fills it. The handle of the cock, *o*, is then drawn into its first position; and the steam again presses the water in the receiver up the pipe *m*.

In Savery's first engines the condensation was invariably accomplished by condensing the vapour, by an affusion of cold water on the outside cylinder, *a*. We have no authentic information at what period the high-pressure engines were constructed to condense by a jet in the inside; probably about 1712. In the engine we are describing, the condensation by injection was improved *by the water being made to fall through a cullender, a, in order, by diffusing the jet more equally, to get a speedier condensation*. It is necessary, however, to remark that this contrivance may not be due to Desaguliers, from his *not* laying claim to it.

Desaguliers tells us he erected seven of these engines in 1717 and 1718. The first was made for Czar Peter I for his Garden at Petersburgh. The boiler was made "spherical, as they must all he in this way where the steam is much stronger than the air, and held between five and six hogsheads; the receiver held one hogshead, and was filled and emptied four times in a minute. The water was drawn up by the suction or pressure of the atmosphere, twenty-nine feet high out of the well, and then pressed up about eleven feet higher. The pipes were all of copper; but soldered to the horse with soft solder, which I knew would hold very well for that height, or a greater height for that quantity; for if the quantity was larger, then the boiler must be greater, and the steam of the same force would have a greater surface to act upon, which might burst the boiler, or require it to be made much thicker."

"Another engine," says the Doctor, "which I put up for a friend about twenty-five years ago, drew up the water twenty-nine feet from the well, and then it was forced up by the pressure of the steam twenty-four feet higher into a cistern, holding about thirty tons, set up at the top of a tower, in order to run down again through a pipe of conduct, and play several jets in the gardens. But sometimes no jets being played, the water was at that height of five or six feet discharged out of the force-pipes, to fill the ponds and water meadows in dry weather; which it dial with a less strength of steam than what drove the water into the tower; or if the same strength were kept up, one aright make eight or nine strokes in a minute, instead of about six, when the water was driven up into the cistern. Upon the safety-valve there was a steelyard, the place of whose weight shows the strength of the steam, and how high it was capable of raising water. But when the weight was at the very end of the steelyard, the steam then being very strong, would lift if up; and go out at the valve rather than damage the boiler. But about three years ago, a man who was entirely ignorant of the nature of the engine, without any instructions undertook to work it; and, having hung the weight at the further end of the steelyard, in order to collect more steam to make his work quicker, hung also a very heavy plumber's iron upon the end of the steel-yard. The consequence proved fatal; for after some time the steam, not being able with the safety-clack to raise up the steel-yard loaded with all this unusual weight, burst the boiler with a great explosion, and killed the poor man, who stood near, with the pieces that flew asunder; there being otherwise no danger, by reason of the safety-valve made to lift up and open upon occasion."

"About as much fire as a common large parlour fire was sufficient to work this engine, and to raise fifteen tons per hour. This engine, according to *my* method, consists of so few parts, that it comes very cheap in proportion to the water that it raises; but it has its limits, as I said before."[55]

It is but justice to Savery to notice the engine which his arch enemy propounds as an improvement, to be only a copy of the mechanism (shown in Figure 12) which the Captain constructed fifteen or sixteen years before. "It has nothing but a nickname to conceal its legitimacy." The mode of forming the valve is taken from Papin, but as applied by the Doctor, it can by no means be considered an improvement.

In comparing his improved engine with the Leaver Engine, often called Newcomen's, the Doctor observes, that "it must not be too small; for then it will have a great deal of friction, in proportion to the water that it raises, and will cost too dear, having as many parts as the largest engines, which are the best and cheapest; in proportion to the water they raise, the friction being always as the diameter; whereas the water raised is as the square of the diameter of the cylinder, and a much greater part of the whole power is employed to move all the little machinery than in a great one. I had an experimental proof of this, at Westminster, in the year 1728 or 1729, when Mr Jones (commonly called Gun Jones) built a king model of the Leaver Engine in my garden; (which model he had a mind to present to the King of Spain). I had at the same time, near the place where he erected his engine, one in Savery's way, which raised ten tons an hour about thirty-eight feet high. He made his boiler the exact size of mine, and his cylinder was six inches bore and about two feet in length. When his model or Leaver Engine was finished, it raised but four tons an hour into the same cistern as mine. It cost him £300; and mine, having copper pipes, cost me but £80."[56]

In the earliest application of Steam, it acted by its expansive power, against the pressure of the atmosphere, as in Decaus's and Savery's Engines, and in one of Papin's inventions it was made to raise a piston. Leupold, the well-known author of the *Theatrum Machinarum*, following this idea about 1720, constructed what may be considered as the first HIGH-PRESSURE LEVER ENGINE, in which the steam was permitted to escape into the atmosphere, after had performed the office of raising pistons attached by-rods to a lever. With a candour unusual in the history of the Steam Engine, Leupold[57] ascribes the sole merit of his contrivance to Dr Papin, as it was to him, he confesses, that he was indebted for the idea of employing the elastic force of steam to raise water; and also because the *four-way* cock was taken from Papin, who had employed it in his Air Engine.

Figure 21 is a view of Leupold's machine, as given by himself. The boiler, *a*, communicates by a cock, *x*, with the bottom of two cylinders, having pistons, *c, d,* moving in them; these pistons are attached to the levers, *g,* and *h*, by the rods, *e,* and *f.* To the other end of these levers are fixed pump-rods, *k, 1*, having pistons at their lower ends working into the barrels or cylinders, *o, p; q*, an eduction-pipe, communicating with the pump, in which the water is forced. upwards; *z*, the fire-place; *y*, the ash-pit; *i, i*, the pivots, on which the levers, *g, h*, move; *x*, a cock, so constructed as to shut off the communication between the boiler and either of the cylinders, while it opens a communication with the atmosphere.

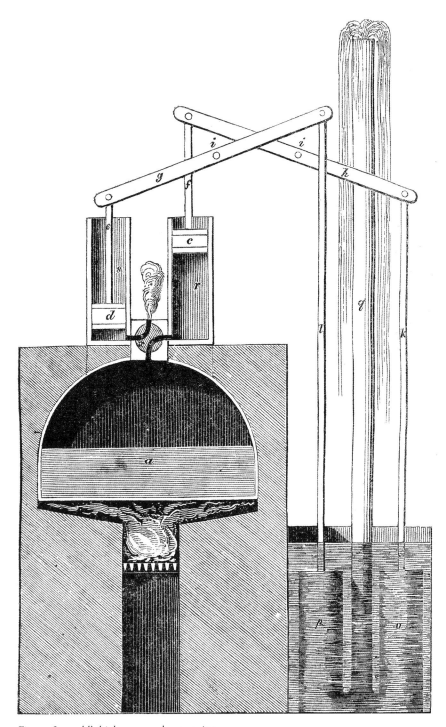

Fig. 21: Leupold's high-pressure lever engine

In the situation of the apparatus shown in the Figure, the steam in the boiler a, flows through the passage 3, in the cock x, into the cylinder r, and presses the piston c, upwards: this depresses the pump-rod k, fixed to the other end of the lever h, which forces the water (under the plunger) up the pipe q. When the steam has raised the piston c, to nearly the top of the cylinder, the cock x is turned, and the passage between the cylinder r and the boiler is closed, and a communication is opened from the inside of the cylinder into the atmosphere. The weight of the rod f, and the piston c, being made greater than k, and o, the piston c falls to the bottom of the cylinder, and the steam which raised it escapes through z, into the atmosphere. This position of the cock is shown in Figure 22.

From the construction of the fourway-cock, at the moment in which the passage of steam into the cylinder r was closed, another passage was opened between the boiler and the cylinder s; the elasticity of the steam forces the piston d upwards, and depresses the plunger at the end of the rod I, and impels the water in the barrel p under it, up the pipe q. When the piston d has reached the top of its cylinder, or made its stroke, the further passage of steam from the boiler is shut off by turning the cock x; and the steam escapes into the atmosphere through z; and d, descends in the cylinder by its preponderance in the same manner as c. During the ascent of d, c has fallen to the bottom of

Fig. 22

the cylinder *r*. The steam-passage from the boiler being then opened, *c* is again raised in its cylinder while the vapour in *s* is escaping into the atmosphere; thus, producing at alternate vertical motion in the pump-rods *k* and *l*. It was subsequently suggested to attach the levers to each other, by which the ascent of one piston depressed the other without employing a counterpoise.

Leupold also proposed using Savery's Engine to raise water by the elasticity of the vapour only. Instead of condensing the steam he allowed it to escape into the atmosphere. *a*, in Figure 23 is the boiler; *b*, a pipe from that into receiver; *c*, the eduction-pipe. When the water has been expelled by the elasticity of the steam, the fourway-cock *e*, shuts off the communication from the boiler, and allows the steam in the cylinder to flow into the atmosphere. It was suggested that by using an intermediate vessel into which the steam might be admitted before escaping, some of the heat might be saved. When the steam has forced all the water from the receiver, a communication is opened into the second vessel, and the cock from the boiler is shut. When the receiver is again filled with water, or while it is filling, the cock is opened, the steam imparts its heat to its sides and to the surface of the water; all the heat, so acquired, is obviously saved; as the same quantity of steam would be condensed, at its admission from the boiler, by the receiver and surface of the water. The

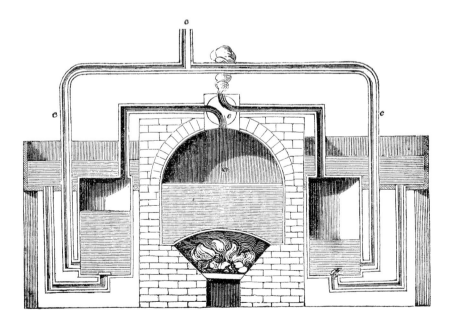

Fig. 23: Leupold's modification of Savery's engine

air which might accumulate in the second vessel was got rid of by a small cock opening into the atmosphere. A separate vessel for this purpose, in Savery's arrangement of the double cylinders when employed only for forcing, was sometimes used; the effect could not be much greater than when an injection was introduced in the usual manner.

MM. Mey and Meyer's apparatus, a description and engraving of which were published[58] by the French Academy of Sciences in 1726, offers no peculiarity in its arrangement or construction, from that shown in Figure 19.

Although Mr Jonathan Hulls did not originate any novelty in the construction of the Atmospheric Engine, he is entitled to the honourable notice of having proposed[59] the application of paddle-wheels moved by a Steam Engine, to propel ships, instead of wind and sails. *In this scheme it was necessary to convert the alternate rectilineal motion of a piston-rod into a continuous rotatory one*, and which he ingeniously suggested might be accomplished by means of a crank. This is now with justice considered to be that invention which introduced the Steam Engine as a first mover of *every variety of machinery*.[60] Hulls was unable to interest the public in his project; and his mode of applying the crank was so completely forgotten, that at its revival, about forty years after this period, a patent was obtained for the invention, and the merit of the application was also claimed by the celebrated Mr Watt, evidently without any knowledge of Hull's suggestion.

The success of the improvements which were gradually introduced into the hand-gear, or self-acting mechanism of the Atmospheric Engines, prompted many schemes to adapt a similar apparatus to those on Savery's construction.

A contrivance described by Gensanne,[61] is shown in Figure 24. x, is the receiver; h, a pipe rising from the cistern; g, the eduction-pipe; f, the injection-pipe; c, lever of injection-cock; b, the axis of the steam-valve, n, which slides horizontally; m, top of boiler; d, d, two tumbling-bobs, performing same office as in the Atmospheric Engine;—the position of the sliding-pin c, is so adjusted, that when the steam-valve is open, the injection-cock is shut. From the manner in which the levers are connected, their simultaneous (rather than self-acting) operation will be so clearly shown by an inspection of the figure, as not to require further illustration. From Gensanne's preliminary observations, it appears that about 1744, several engines had been erected in France. He mentions one at Fresne near Condé; one at Sars, near Charleroi, where it was employed in draining a coal-mine; and a third at the lead-mines near Namur. The celebrated Raider gives a good description of the engine at Fresne;[62] to which place he made several journeys to observe it in operation, while he was drawing up his description.

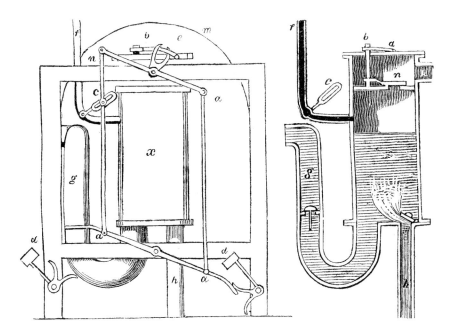

Fig. 24: Gensanne's engine

De Moura, a Portuguese gentleman, also proposed an ingenious arrangement of levers for this purpose; of which he submitted a model to the Royal Society. It is described in their Transactions; but the Atmospheric Engines having very generally superseded Savery's, it did not receive that degree of attention to which, under other circumstances, it would have been entitled. "What is peculiar to this engine, is a float with the receiver, composed of a light ball of copper, which is not loose therein, but fastened to the end of an arm, which is made to rise and fall by the float, while the other end of the arm is fastened to an axis. This axis is made conical, and passes through a conical socket, and by means of the rising and falling of the surface of the water within the receiver, communicates a corresponding motion to the outside, in order to give proper motion to the rest of the gear.[63]"

Nothing that had been done since Beighton's time had any great effect in diminishing the enormous waste of fuel, which took place in the most improved forms of the engine. Different constructions of the fire-place and boilers were proposed as remedies for this evil, but in practice they gave no advantage. Mr Payne's[64] scheme for allowing very small streams of water from revolving pipes to fall on the inside of a hollow cone of cast-iron kept red-hot, was not found to be efficient for the production of the necessary quantity of vapour, and soon destroyed the apparatus. Mr Smeaton conducted the flame through a great length of circuitous channel, before it entered the chimney, with the hope of arresting a greater proportion of the heat by detaining the smoke a longer time under the boiler. On trial it was found that the draught or flow of the air through the fuel was so weakened, that more was lost by the imperfect combustion of fuel from this cause, than was pined by the longer retention of the heated vapour. The theory of combustion had but begun to attract the attention of experimental philosophers; and it was many years after this period before any progress was made in its investigation. Its application to the economy of fuel in furnaces is due to the labours of a very recent period.

The patent obtained by Mr Blakey in 1756 was for preventing the water in Savery's engine from coming into contact with the steam, by *interposing a quantity of oil between them*: this, from its less specific gravity, floating on the surface of the water in the receiver, operated as a non-conducting piston. He likewise employed *two receivers*, one placed over the other; so that the water under the oil would always be the same, instead of being renewed at each injection. *Air* might also be interposed between the steam and the water, to prevent their coming into contact. The steam which was admitted into the highest receiver being lighter than the air, occupied the upper part, and pressing the air before it, displaced an equal quantity of water, by forcing it up the eduction-pipe. In practice this was not found to answer; the compressibility of the air absorbed too great a portion of the power; and besides, although the steam was prevented from coming into contact with the surface of the water, either by employing air, or oil, or both,—still, from the construction of all engines on this principle, the oil and the air must wine into contact with the sides of the receiver, and surface of the water—and lose a portion—a great portion, of their temperature. These, in their ascent into the first receiver, absorbed part of its heat, and operated in the same manner as the water used in the common engines; although, from the lower conducting power of oil or air, their absorption of heat, from the receivers, was less than in engines on the usual form.

This saving was not, however, found to be of so much amount as to overbalance the disadvantages which were experienced by the interposition of the air, particularly, as a moveable piston. "Great contention arose among some of those who counted themselves men of science," says Hornblower, "as to the practicability of such a project: some giving it as their opinion, that if the principle were to be admitted, it would be very difficult to apply it in mines, where it would require ten atmospheres at least while others, with exalted pretensions, declared it possible to conduct its influence to the centre of the earth. But an accident terminated the event as to this engine in Cornwall, by one of the steam-vessels bursting, through the force of the steam, though much under the degree of power proposed to the Cornish gentlemen."[65] Mr Blakey's attempts to introduce it into other countries were probably unsuccessful from similar failures. "Such is the degeneracy of man," continues Hornblower, "that while the Academy of Sciences at Paris, and the delegates of the States-general in Holland, were pluming him with the gaudiest expressions of their approbation, not one instance can be found where he received the encouragement he was led to expect."[66]

Blakey's Engine is shown in Figure 25. E, is the boiler; D, a pipe proceeding from it into the receiver, or air-vessel, I: this receiver is placed over another vessel V, of equal capacity, and joined to it by the pipe f; the receiver V, is connected by the pipe Q, with the well suction-pipe and eduction-pipe X; C, is a funnel to supply the boiler with cold water; T, the injection-pipe and cock L; P, a cock to supply air to the receiver, or to permit its escape when its accumulation renders it necessary: this is also employed as a *gauge-cock*; F, a gauge-cock on the boiler. The fire-place, ash-hole, and chimney, are similar to those of all other engines we have described.

The operation of this engine is the same as Savery's, and abundantly simple. Steam is generated in E, which flowing through D, displaces the air in I, (driving it before it like a piston) through f, into the receiver V; which pressing on the water in V, forces it through Q, up the eduction-pipe X; when all the water is expelled from V, up the pipe X, into the reservoir, the injection-cock L is opened, and the water falling through the small holes in the cub lender at the end of pipe T, condenses the steam in I, and falling through a second cullender placed in the upper part of V, it also condenses the steam in that vessel. A vacuum being produced in I and part of V, the pressure of the atmosphere forces the water in the well m, up the pipe n, into the receiver V. When it arrives at the height of S, the steam-cock D, is opened, and the vapour from the boiler entering the vessel I, presses the air it contains through f into V, and displaces the water in V, which makes it flow through Q into X. When the vessel V is emptied, the steam-cock is shut, and the injection-cock opened, and the pressure of the atmosphere raises the water from M, again into V. This, it will be perceived, is only Savery's Single Engine; for no benefit is obtained by two receivers connected by a pipe. In Savery's the interposition of the air, as a piston, between the steam and the water, was considered a practical defect, for which there could only be an

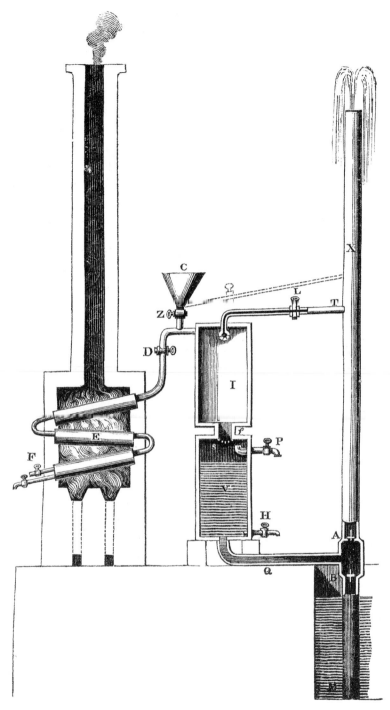

Fig. 25: Blakey's engine

occasional remedy: and it was not surprising that Blakey's scheme to make use of this defect, with a prodigious extension, should fail in practice.

In the engraving the cocks are shown as if formed to be opened and shut by hand; but in practice Blakey adapted a series of levers to perform these operations; and their alternations were regulated by a hollow sphere which rose and fell with the surface of the water, in the manner of a common ball-cock. To generate steam of the required elasticity, and in sufficient quantity, he contrived a boiler, which is shown in the engraving, and will be clearly understood from inspection.[67] It was not, however, included in his patent, nor invented for many years after it: its place as to date is about 1774. We insert it here to give his improvements in one engraving. It is on the same principle as that recently introduced by Woolf, and employed with great advantage when attached to his engines.

In 1758, Mr Dean Fitzgerald published in the *Philosophical Transactions*, a method of converting the reciprocating motion of the Atmospheric Engine into a rotatory one, "by employing *a combination of large toothed wheels, and of smaller ratchet wheels worked by teeth upon the arch or sector of the great beam*: one of these ratchet wheels being put in motion by the ascent of the beam, and standing still during its descent, when another ratchet wheel is moved by an intervening wheel in the same direction as the first; and thus the two communicate a continuous rotative motion to the axis on which they are placed, which is thence transmitted by a larger toothed wheel to a smaller wheel or pinion, on the shaft of which is *a fly to accumulate momentum*, and a crank proposed to be applied to work ventilators and to many other useful purposes. *The fly, by accumulating in itself the power of the machine during the time it was acted upon, would continue in motion, and urge forward the machinery whilst the Steam Engine was going through its inactive returning stroke.*" It would seem that Mr Fitzgerald had made a model of his contrivance; but it did not appear to Mr Watt, who had made an accurate search at the office, that any patent was ever taken out for the invention. "Fitzgerald published proposals for erecting *mills of all kinds* driven by Steam Engines; but these proposals do not appear to have met with the confidence of the public."[68]

Savery's Engine had fallen into disuse, and after Smeaton had pointed out the principles which should govern the construction and proportion of the parts of the Atmospheric Engine, little scope for improvement appeared to be left for, other mechanics.

The celebrated Brindley, in 1759, attempted to reduce the quantity of coals used by Steam Engines, by forming the sides and top of the boiler of wood, and the bottom of stone. The fireplace was made in the inside of the boiler, and was surrounded with water. But this scheme was abandoned, in consequence of the rapid destruction of the wood by the steam; and Brindley resorted to the practice of casing iron boilers with wood.[69] Smeaton at this time began to form the under side of the steam piston of the same material, and with hardly greater advantage.

Many engines on Newcomen's arrangement had been erected in various parts of France, and were employed in three instances in Holland. Blakey also erected some engines on Savery's plan for the purpose of draining land. About 1760, the Atmospheric Engine was in use in the British American States, but the purpose to which it was applied is not stated. In the province of New England there were two in operation a little after this period, one of them for draining a copper-mine on the Passaick.[70]—Those which were in action in Holland and France, like those in America, had been manufactured in England.

The increasing commerce of the country had given a new impulse, and in numerous instances a more beneficial direction, to those operations connected with the construction of machines in general. The woollen manufacture, enlarging its boundaries, created a demand for the labours of the machinist, and opened a tempting and flattering prospect for mechanical inventions, to abridge or to expedite its manipulations. The cotton trade, which has in our times received such a prodigious extension, was at this period establishing itself; and, from the novelty of its processes and importance of its fabric, eminently assisted to stimulate mechanical skill and invention. The Atmospheric Engine, employed extensively as a drainer of mines, having begun to be considered, since Fitzgerald's proposal, as offering a power capable of being applied as a general first mover,[71] also tended to introduce among artisans a feeling of the necessity of more correct and varied knowledge than was demanded from them when water or windmills, or animal labour, were exclusively resorted to, to give motion to pumps and to the few other machines which then sufficed for the manufacture of those commodities, which, from local circumstances favourable to their production, were considered as the staple produce of particular districts.

Availing themselves of this increasing mechanical taste, several enterprising individuals had opened a field of honour and emolument, by reading lectures in the provincial towns, on the more attractive branches of mechanical philosophy; the resident professors in some of the colleges following the spirit of the times, had also begun to adopt a more popular manner of imparting those principles of science which are more immediately applicable to mechanical purposes, than had hitherto been practised in those learned establishments. In the enlightened endeavour to illustrate the truths of philosophy by the operations of the manufacturer, that constellation of great men who, about the middle of the eighteenth century, filled the chairs of the various professorships in the college of Glasgow, were honourably distinguished; and to their exertions in diffusing a taste for blending science with usefulness, was mainly owing the opening of the channel of that prosperity, which has since flowed in a full and tideless stream upon their beautiful city—now so distinguished for the number, enterprise, and honour of its merchants—so pre-eminent for the intelligence, love of science, and admirable mechanical skill of its artisans.

The museum of this distinguished body was unusually complete in its collection of mechanical apparatus, and, above all, it possessed a small working model of a Steam Engine, a circumstance memorable among mechanics for the important results which its possession occasioned.

We have, in the course of this essay, had many occasions of mentioning the name of Dr Robison as an author of the highest—of the very first authority, in the History of the Steam Engine. The treatises inserted in the *Encyclopaedia Britannica* did him infinite honour; and he may be instanced as among the few English mechanical writers possessed of the rare talent of imparting his own varied and profound acquirements, in a style which delights the scholar by its animated elegance, and allures the mechanic by its perspicuity and simplicity. We have thought it our duty, therefore, to point out some errors in the Doctor's work, solely from the circumstance of the authority of his name being so great as to have given currency to statements which have misled authors whose acquaintance with the subject, and opportunities of observation, ought to have enabled them to have been more accurate.

At the date of our narrative, he was a very young man studying at the University of Glasgow, devoted to mechanical pursuits; and in habits of intimacy with Mr James Watt, who had lately been appointed to the care of the university's collection of mechanical and philosophical models, and then living in apartments within the college. At this period Robison had anticipated an extension of the use of the Steam Engine to some purposes which have been achieved by succeeding mechanics;—but independently of this merit, as well as his being the best historian of the progress of the improvements made on the Engine, he celebrated as having first drawn the notice of his friend Young[72] to the improvement of steam apparatus.

"My attention," says Mr Watt,[73] "was first directed in 1759 to the subject of Steam Engines by Dr Robison, then a student in the university of Glasgow, and nearly of my own age. Robison at that time threw out the idea of applying the power of the Steam Engine to the moving of wheel carriages, and to other purposes; but the scheme was not matured, and was soon abandoned on his going abroad.

"In 1761 or 1762, I made some experiments on the force of steam in a Papin's digester, and formed a species of Steam Engine, by fixing upon it a syringe one third of an inch in diameter, with a solid piston, and furnished also with a cock to admit the steam from the digester, or shut it off at pleasure, as well as to open a communication from the inside of the syringe to the open air, by which the steam contained in the syringe might escape. When the communication between the syringe and digester was opened, the steam entered the syringe, and by its action upon the piston raised a considerable weight, (15 lb.) with which it was loaded. When this was raised as high as was thought proper, the communication with the digester was shut, and that with the atmosphere opened; the steam then made its escape, and the weight descended. The operations were repeated, and, though in this experiment the cock was turned by hand, it was easy to see how it could be done by the machine itself, and to make it work with perfect regularity. But I soon relinquished the idea of constructing an engine upon this principle, from being sensible it would be liable to some of the objections against Savery's Engine; for the danger of bursting the boiler and the difficulty of making he joints tight; and also that a great part of the power of the steam would be lost, because no vacuum was formed to assist the descent of the piston."[74]

The attention necessary to his business of a mathematical instrument maker,[75] prevented Mr Watt from prosecuting the subject any further at this time. But in the winter of 1763 and 4, having occasion to repair a model of Newcomen's Engine, belonging to the Natural Philosophy Class of the University, his mind was again directed to the subject. "At that period," he informs us, "his knowledge was derived principally from Desaguliers, and partly from Belidor. He set about repairing the model, *as a mere mechanician*; and when that was done and set to work, he was surprised to find that its boiler was not supplied with steam, though apparently quite large enough (the cylinder of the model being two inches in diameter, and six inches stroke, and the boiler about nine inches in diameter): by blowing the fire it was made to take a few strokes; but required an enormous quantity of injection-water, though it was very lightly loaded by the column of water in the pump. It soon occurred to him that this was caused by the little cylinder exposing a greater surface to condense the steam than the cylinders of larger engines did, in proportion to their respective contents; and it was found that by shortening the column of water, the boiler could supply the cylinder with steam, and that the engine would work regularly with a moderate quantity of injection. It now appeared that the cylinder being of brass, would conduct heat much better than the cast-iron cylinders of, larger engine's (which were generally lined with a stony crust), and that considerable advantage could be gained by making the cylinders of some substance that would receive and give out heat the slowest. A small cylinder, of six inches diameter and twelve inches stroke, was constructed of wood[76] previously soaked in linseed oil, and baked to dryness. Some experiments were made with it; but it was found that cylinders of wood were not at all likely to prove durable; and that the steam which was condensed in filling it, still exceeded the proportion of that which was required for engines of larger dimensions. It was also ascertained that unless the temperature of the cylinder itself were reduced as low as that of the vacuum, it would produce vapour of a temperature sufficient to resist part of the pressure of the atmosphere. All attempts therefore to produce a better exhaustion, by throwing in a greater quantity of injection water, was a waste of steam, for the larger quantities of injection cooled the cylinder so much, as to require quantities of steam to heat it again, out of proportion to the power gained by having made a more perfect vacuum; and on this account the old engineers acted wisely in loading the engine with only six or seven pounds weight on each square inch of the piston.[77]"

By subsequent experiments Mr Watt[78] also ascertained that steam was about 1800 times rarer than water. In another experiment, being astonished at the *quantity* of water required for the *injection,* and the *great heat* that it had acquired from the small quantity of water in the form of *steam,* which had been used in filling the cylinder; and not understanding the reason of it, "I mentioned it," he says, "to my friend Dr Black, who then explained to me his doctrine of *latent heat,* which be had taught for some time before this period (summer 1764); but having myself been occupied with pursuits of business, if I had heard of it, I had not attended to it, *when I thus stumbled upon one of the material facts upon which that beautiful theory is founded.*"[79]

On reflecting further, it appeared to him that in order to obtain the greatest power from the steam, the cylinder should always be kept as hot as the steam which entered it; that when the steam was condensed, the water of condensation and the water of injection should be cooled to 100 degrees Fahrenheit, or lower if possible.

The means for accomplishing these two grand objects did not present themselves to Mr Watt at the moment when he had drawn those sagacious inferences; but it occurred to him early in the year 1765,

that if a communication were opened between a cylinder containing steam, and another vessel which was exhausted of air and other fluids; the steam, as an expansible fluid, would immediately rush into the empty vessel, and continue to do so until it had established an equilibrium: and if that vessel were kept very cool by an injection or otherwise, more steam would continue to enter, until the whole was condensed.

Admirable Invention!

Thus was accomplished what had been considered impossible by all previous engineers,—*the production of a vacuum* without *cooling the cylinder*.[80]

But if both of the vessels be exhausted, or nearly so, how were the injection-water, and the air entering with it, and also that produced by the condensation of the steam, to be extracted from them? This, Mr Watt proposed to do by adapting to the condenser, a pipe, whose length would exceed that of the length of a column of water equivalent to, the pressure of the atmosphere, and to extract the air by means of a pump—or to employ a pump to extract both the water and the air.

Instead of keeping the piston tight by water, which could not be applicable in this new method, as, if any of it entered into a partially exhausted and (now) *hot cylinder*, it would boil, and by generating vapour, prevent the production of a vacuum, besides cooling the cylinder, by its evaporation during the descent of the piston,—*he proposed to lubricate the sides and keep the piston air-tight, by employing wax or tallow.*

It next occurred to Mr Watt, that, the mouth of the cylinder being open, the air which entered to act on the piston would cool the cylinder, and condense some steam on again filling it. Then he proposed, "to put an air-tight cover on the cylinder, with a hole and stuffing-box for the piston to slide through, and to admit steam above the piston, to act upon it instead of the atmosphere."

This was his second grand improvement; and while the power of the mechanism remained untouched, the expense of fuel or waste of steam was reduced to nearly a third of its former amount; and the machine was now, properly, an engine, acting by the force of *steam*—the motive power being derived hitherto from the gravity of the atmosphere.

The other source of the loss of heat, by the air of the atmosphere cooling the cylinder externally, which produced a condensation of then internal steam; was obviated in thought *by enclosing the steam cylinder in another of wood, or of some other substance, which would conduct heat slowly.*

When once the idea of separate condensation was started, all these improvements, continues this admirable mechanic, were suggested in quick succession; so that, in the course of *one or two days*, the *invention was so far complete in his mind*, and he immediately began to submit them to the test of experiment.

The model he used consisted of a brass syringe 1¾ inch in diameter, and ten inches long; having a cover at top and bottom of tin plate, a pipe to convey steam to both ends of the cylinder from the bottom, and a second pipe to convey the steam from the upper end of the cylinder to the vessel in which the steam was to be condense. To save apparatus the cylinder was inverted. A hole was drilled longitudinally through the axis of the stem of the piston, and a valve was fixed at its lower end, to permit the water produced by the steam which was condensed at first entering the cylinder to escape. The *condenser* was made of two pipes of tin plate ten or twelve inches long, and about a sixth of an inch in diameter, placed perpendicularly, and communicating at top with a short horizontal pipe of large diameter, shut at its upper end, with a valve opening upwards. These pipes were joined at bottom to another perpendicular pipe about an inch in diameter, which served for the air and water-pump. These pipes and pump were all placed in a cistern filled with cold water.[81]

The steam-pipe was attached to a small boiler. When steam was generated; it filled the cylinder, and soon issued at the longitudinal perforation of the rod, and through the valve into the condenser: when it was judged the air was expelled, the steam-cock was shut and the air-pump piston-rod was drawn up, which leaving a vacuum in the condenser pipes, the steam entered them and was condensed: the piston of the cylinder immediately rose and lifted a weight of about eighteen pounds, which was hung to the lower end of the piston-rod. The exhaustion-cock was shut, the steam was readmitted into the cylinder, and the operation was repeated; and excepting the non-application of the steam-case, and external covering to prevent the dissipation of the heat by radiation, the invention was complete as far as regarded the savings of steam and fuel. To verify the expectations that Mr Watt had formed of the advantages of his invention, he constructed a large model with an outer cylinder and wooden case, the effect of which exceeded his most sanguine expectations.

This form of the condenser was afterwards changed; it being found that to condense the steam used in a large engine by the cold water being applied on the outside of the condenser, would require vessels of large and very inconvenient dimension. And it was also found that, from the nature of the water with which the engines are frequently supplied, a stony crust was quickly formed upon the outside of the iron plate of the condenser, which greatly diminished, or altogether destroyed its conducting power.

About the period of his marriage with Miss Miller in 1764, Mr Watt removed from his apartments in the College, and commenced practising as a land-surveyor, by the advice and with the occasional assistance of his uncle. He soon got into a respectable practice, and was frequently consulted on the formation and improvements of canals and harbours, some of which have since been executed; but whether it was from his time being completely occupied with the duties of his new profession, "or the indifferent state of his health, or his want of funds, or his apprehension of the prejudices and opposition that he might have to encounter,"[82] which prevented him from securing his invention by a patent, we know not—yet, although fully sensible of the value of his discovery, he proceeded no further in it at this time. With a felicity of invention and powers of mind which rarely fall to the lot of man, so retiring and reserved were his habits and manners, that his diffidence prevented a disclosure of his great discovery even to his friends: indeed at no period of his life was he possessed of that self-confidence which was necessary to bring him or his inventions before those whose patronage and assistance might be required to enable him to carry his designs into execution.

In 1765, Smeaton[83] had constructed a portable Atmospheric Engine, and employed it in draining. Two years after this period some attempts were made by Mr John Stewart,[84] and Mr Dugald Clarke, to produce a continued rotative motion from Newcomen's Engine, to be applied to some sugar-mills in Jamaica; but from their being complex and liable to breakages, they were abandoned. "About 1768 an Atmospheric Engine had been employed at Hartley Colliery, to draw coals from a pit, which had a toothed sector on the end of the working-beam, working into a trundle, which, by means of two pinions with ratchet-wheels, produced a rotative motion in the same direction, by both the ascending and descending stroke of the arch; and by shifting the ratchets, the motion could be reversed at pleasure. The engine had no fly-wheel, and went sluggishly and irregularly."[85]

In the course of his employment of a land-surveyor, Mr Watt had formed an intimacy with Dr Roebuck, an English gentleman (the founder of the Carron iron-works,) of great enterprize, and some fortune, who had embarked in extensive speculations in coal-mining and making salt, at Borrowstoness in the county of Linlithgow. By the Doctor's assistance Mr

Watt was enabled to erect one of his Engines at a coal-mine on the estate of the Duke of Hamilton, at Kinneil, about a mile from Borrowstoness.. This had a cylinder of eighteen inches; "and being a sort of experimental Engine, was successively altered and improved till it was brought to considerable perfection."[86] During its erection, in the winter of 1768, he applied for a patent to secure the profits of his. Ingenuity, which was enrolled in April 1769. On a subject so important, it cannot be superfluous to insert the words of the inventor, whose admirable application of the sciences to practical purposes most justly entitles him to a rank among philosophical mechanics, not inferior to that of Ctesibius and Dr Hooke.[87]

"My method of lessening the consumption of steam, and consequently fuel in fire engines, consists in the following principles:—First, that the vessel in which the powers of steam are to be employed to work the engine, which is called the cylinder in common fire-engines, and which I call the steam vessel, must, during the whole time the engine is at work, be kept as hot as the steam which enters it; first, by enclosing it in a case of wood, or any other materials that transmit heat slowly; secondly, by surrounding it with steam or other heated bodies; and thirdly, by suffering neither water, or other substance colder than the steam, to enter or touch it during that time. Secondly, in engines that are to be worked wholly or partially by condensation of steam; the steam is to be condensed in vessels distinct from the steam vessel or cylinders, though occasionally communicating with them. These vessels I call *condensers*, and whilst the engines are working, these condensers ought at least to be kept as cold as the air in the neighbourhood of the engines, by application of water or other cold bodies. Thirdly, whatever air or other elastic vapour is not condensed by the cold of the condenser, and may impede the working of the engine, is to be drawn out of the steam vessels or condensers by means of pumps, wrought by the engines themselves, or otherwise. Fourthly, I intend in many cases to employ the *expansive force of steam to press on the pistons*, or whatever may be used instead of them, in the same manner as the pressure of the atmosphere is now employed in common fire-engines. In cases where cold water cannot be had in plenty, the engines may be wrought by this *force of steam only* by discharging the steam into the open air, after it has done its office. Fifthly, where motions round an axis[88] are required, I make *the steam vessels in form of hollow rings, or circular channels, with proper inlets and outlets, for the steam, mounted on horizontal axles like the wheels of a water-mill.* Within them are placed a number of valves, that suffer any body to go round the channel in one direction only: in these steam vessels are placed weights, so fitted to them as entirely to fill up a part or portion of their channels, yet rendered capable of moving freely in them by the means hereinafter mentioned or specified. When the steam is admitted in these engines between these weights and the valves, it acts equally on both, so as to raise the weight to one side of the wheel, and by the reaction on the valves successively to give a circular motion to the wheel; the valves opening in the direction in which the weights are pressed, but not in the contrary. As the steam vessel moves round, it is supplied with steam from the boiler, and that which has performed its office may either be discharged, by means of condensers, or into the open air. Sixthly, I intend in some cases to apply *a degree of cold not capable of reducing the steam to water, but of contracting it considerably, so that the engines shall be worked by the alternate expansion and contraction of the steam.* Lastly, instead of using water to render the piston or other parts of the engines air or steam; tight, I employ *oils, wax, resinous bodies, fat of animals, quicksilver, and other metals in their fluid state.*"[89]

These improvements were combined in a very masterly manner, in what were called his single reciprocating Engines. The lever-beam, boiler, pump-rod, and the use of a plug-frame to act by its pins or tappets on the hand-gear, or contrivances for opening and shutting the valves, with some improvements in their arrangement; were retained; the valves were, however, on a different and a better construction.

In Figure 26 the pipe *d,* conducts steam from a boiler (which is omitted in order to have the principal parts of the apparatus on a larger scale); *e,* is the nozzle, or square box; containing a valve, which in its rise or fall opens or shuts the passage between one side of the piston and the boiler, and also between the pipe *o, 1,* and the cylinder *a.* Y Y is the interstice between the casing and its cylinder: the casing was called the *jacket.* This interval was sometimes filled with charcoal, or some other slow conducting substance; or steam from the boiler might be admitted into it: by any of these means the radiation of heat from the steam cylinder, and its conduction by the air, was very perfectly prevented. *b,* is the steam piston attached to the lever-beam; by the rod *x.* The pins or tappets *n, n,* fixed, on the *plug-frame* (or tappet rod,) which in our Figure also serves for the rod of the pump, attached to the condensing apparatus: at the ascent or descent of these pins, they strike on the ends of the levers or spanners *m, m, m,* connected with the valves, *e, f, c,* and open or shut them, as they may be adjusted. The condenser *h,* is connected with the steam cylinder by the pipes *u,* and *g.* The air-pump barrel is attached to this vessel by the pipe *s,* which is furnished with a valve opening from the condenser: the piston is similar to those usually employed in water-pumps, with the exception of the joints of the valves being made of metal instead of leather; the

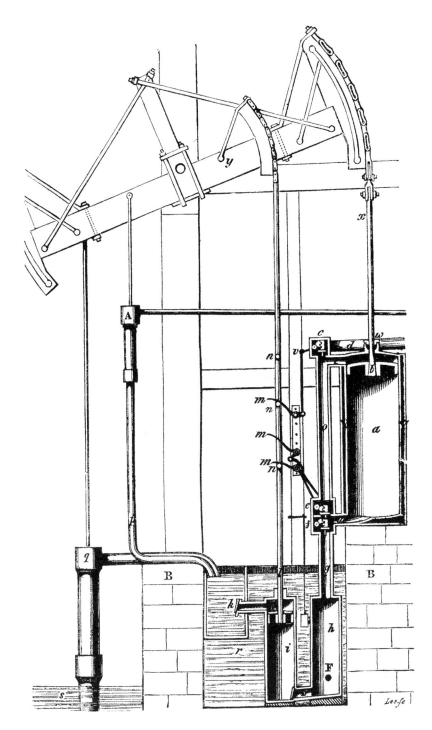

Fig. 26: Mr Watt's single engine

condensing pump, *i*, has a short pipe proceeding from near its top, on the end of which is, a valve, *k*, opening outwards, and into a vessel of water, *r*: the condenser and its pump are placed in a cistern of water, kept as cold as possible, by allowing the heated water to escape, and renewing the supply from the mine or some reservoir by a pump, *q*, which is worked by a rod attached to the lever-beam. The short pipe k_2 opens into a cistern generally separated from that in which the condenser is placed, in the manner shown in the Figure. From this cistern water is pumped into the boiler through the pipe *p*, by the pump A, to supply its waste from evaporation. E, is a foundation of masonry or wood, on which the cylinder is fixed; *o, a*, post or beam to receive the spanner fulcrum—in some engines it is supplied by brackets fixed to the cylinder; *w*, is a *stuffing-box*, first used by Mr Watt, to keep the aperture in which the piston-rod slides steam-tight; the rod of the mine pump is suspended at the opposite end of the lever to the steam piston-rod, as in the Figures of the Atmospheric Engine; (the limits of our page not permitting a greater extension, it is not shown in the engraving). This end of the lever, or the rods attached to it, are made so much heavier than those on the other side of the axis, as to be sufficient to act as a counterpoise to raise the steam-piston from the bottom to the top of the cylinder. The steam-cylinder is closed at top, and the rod slides through the stuffing-box; so that all communication with the atmosphere, and any part either of the cylinder or the condenser, is completely prevented.

If we now suppose the apparatus to be in the position shown in the Figure, and a proper supply of steam in the boiler, the valves *c, e, f,* are to be opened, and N is shut. Steam then enters above and below the piston into the pipe *o*, and the condenser *h*; when the cylinder becomes sufficiently heated, the steam will descend into the condenser, and from its less gravity occupying the upper portion, will expel all the air through the blowing valve, *f*; which may have collected in those vessels, or which may have been produced by the condensation of the steam in heating the apparatus. The cock *e*, is now to be shut, and the injection-cock opened; the jet of cold water will condense the steam in the condenser, and the steam under the piston in the cylinder, *a*, rushing through the pipe *u, g*, to restore the equilibrium, will also be condensed by the jet which is kept playing; and a vacuum is formed in the cylinder and condenser. This condensation in the largest engines is prodigiously rapid; in practice it may be considered quite instantaneous. The communication between the boiler and the upper side of the piston remaining open, the elasticity of the vapour having no now resistance from that on the other side, presses the piston downwards into the vacuous space until it reaches the bottom of the cylinder. At this instant the descent of the tappets in the plug-frame strikes the end of the spanners m, m, connected with the valves c, f, which depresses them, and prevents the farther flow of steam from the boiler, and also shuts off the communication between the condenser and the cylinder.

The piston of the condensing-pump, being attached to the same side of the axis with the cylinder, is also at the bottom of the barrel, and all the water and air which it contained having lifted up the valves which only rise upwards, is now above its piston, the valve s, opening from the condenser, preventing its return into that vessel.

It now becomes necessary to raise the steam piston to the top of the cylinder.

The mine-pump end of the lever-beam, and its rods, have been stated to be made heavier than those of the steam piston and condensing pump, in order to act as a counterpoise to raise the steam piston.

But before this counterpoise can be brought to act, all the steam and air which has just depressed the piston, and still remains above it, must be got rid of. Mr Watt accomplished this desideratum by a contrivance so simple, and yet so refined, as almost to equal in ingenuity the application of the condenser itself. He connected the top and bottom of the cylinder by a pipe, in which he placed a valve, c: when, therefore, the valves e, and f, were closed, this valve was opened; the steam which was above the piston was now admitted under it, and the counterpoise raised the piston in a non-resisting medium.

The opening of this valve, c, is but a momentary operation, and the rise of the piston is not perceptibly of longer duration than its descent. When the counterpoise has raised it to the assigned height, a tappet acts on the end of the spanner connected with the valve e, and closes it, while at the same moment other tappets act on their levers to open again cocks c and f, which are attached to them.

The steam which had flowed through o and e, from the top of the cylinder to the space at the under the piston, now rushes through u and g, (as at the first descent of the piston,) and a vacuum is instantaneously formed in the cylinder, which allows the expansive force of the steam to press the piston downwards, until it has reached a second time the limit of its stroke. The tappets and spanners now close c and f, and open e, when the counterpoise again operates to draw up the piston.—The counterpoise has been stated as being made equal to raising the steam piston, but in fact it must also, in this arrangement of the parts, elevate the water and the condenser pump.

The operation of the condenser pump is very simple. The water which spouts into the condenser through the injection-cock N, increased by the very small quantity arising from condensation, falls to the bottom of the condenser *h*. The upward motion of the pump-rod forming a vacuum in *i*, opens the valve *s*, which allows the water to flow through it from *h*: the water lying on the top of the piston being expelled through *k*, into the cistern. The descent of the piston pressing on the water at the bottom of the barrel, which is prevented from returning, by its opening inwards, the water and air raise the valves, which falling at the moment of the ascent of the piston, raise whatever water and air is above them until they are expelled, at *k*. The hot water from this pump is generally again introduced into the boiler.

In Mr Watt's first trials with the condenser, he procured a vacuum by cooling the condenser externally, as in the first Atmospheric Engines: but the same objection applied to his experiments as to Newcomen's—the condensation was very imperfect. He next *introduced a jet into the pipe connecting the cylinder and the condenser*; and in the progress of his improvements it was introduced in the manner shown in our Figure. The injection-cock is furnished with a lever, for the purpose of enlarging or diminishing the aperture by which the jet is formed, when a greater or less quantity of injection water is required for the purpose of condensation, and which will obviously vary, either from the temperature of the water which is admitted, or the elasticity of the vapour which may be condensed. This lever and its handle are shown between *e* and *f*; and N is the projection or box containing the cock.

It had been customary among engineers before Mr Watt's time, to form the lever-beam by bolting a great many solid beams together, to obtain the necessary stiffness. Some of Smeaton's were constructed with considerable science, but on account of their great expense, many of them costing £800 or £900 and their enormous weight and consequently cumbrous proportion, were superseded by others, in which stiffness was obtained by a mode of trussing with iron rods instead of using a mass of materials. This improvement will be more apparent by comparing the engraving of Smeaton's lever-beam, which he constructed for the engine he erected for the Chacewater Mine, with that on the Figure before us.[90]

In the course of Mr Watt's experiments, even before he procured his patent, he had been struck with the remarkable fact of what is now called the expansion of steam when admitted into a vacuum: and in a letter of his to Dr Small of Birmingham, dated at Glasgow in May 1769, he proposed to avail himself of this property to increase the effect of his engine, and to save steam. "I mentioned to you a method of still doubling the effect of the steam, and that tolerably easy, by using the power of *the steam rushing into a vacuum, at present lost*. This would do little more than double the effect, but it would too much enlarge the vessels to use it all. It is peculiarly applicable to wheel-engines, and may supply the want of a condenser where force of steam only is used; for, open one of the steam-valves and admit steam until one fourth of the distance between it and the next valve is filled with steam, shut the valve, and the steam will continue to expand and press round the wheel, with a diminishing power, ending in one fourth of its first exertion. The sum of the series you will find greater than one half, though only one fourth of steam was used." The sentence which follows this is also remarkable: "The power will indeed be unequal, but this can be remedied by a fly, or several other ways."[91]

After erecting the engine at Kinneil, Mr Watt had begun to make arrangements to manufacture his Engines on a considerable scale, when, the failure of some of the great mining speculations in which his partner, Dr Roebuck, had embarked, producing pecuniary embarrassment, and the Doctor being disabled from rendering further assistance, by making the stipulated disbursements, Mr Watt was on the eve of abandoning the further prosecution of his project. With his consent a negotiation was opened with Mr Matthew Bolton of Birmingham, a gentleman who some years before had founded an establishment at that place, and had already become known as one of the most intelligent and enterprising manufacturers in the kingdom. The negotiation was concluded in 1773, when Dr Roebuck resigned his share of the patent to Mr Bolton, upon terms deservedly most advantageous to himself.

A brighter prospect now opened on Mr Watt; his new colleague was a man of affluence and of great personal influence, "and to a most generous and ardent mind, he added an uncommon spirit for undertaking what was great and difficult. Mr Watt was studious and reserved, keeping aloof from the world; while Mr Bolton was a man of address, delighting in society, active, and mixing with people of all ranks with great freedom, and without ceremony. Had Mr Watt searched all Europe, he could not have found another person so fitted to bring his invention before the public in a manner worthy of its merit and importance; and although of most opposite habits, it fortunately so happened, that no two men ever more cordially agreed in their intercourse with each other."[92]

At this time Mr Wyatt removed to Birmingham, and his first operations were directed to the construction of an Engine, which was erected at Soho, for the inspection of those engaged in mining. But, finding that the period of the patent must expire, before he could possibly be reimbursed for the, mere expenses which must be incurred, before he could hope to complete the necessary arrangements and machinery for the manufacture of engines of such magnitude and demanding so much accuracy in their construction;

about the end of 1774, by the aid and advice of Mr Bolton, and the zealous co-operation of Dr Roebuck and some philosophical friends, who generously gave him the influence of their favourable opinion of the merit of his invention; he was encouraged to apply to Government to extend the term of his patent. Early in 1775, he succeeded in obtaining an Act of Parliament, granting him an exclusive privilege to manufacture his improved Engines for the term of twenty-five years from the date of his application.

Mr Bolton[93] now became his partner in the manufacture of his engines, and a part of the establishment at Soho being appropriated to this purpose; "with his assistance in systematizing the manufacture of the parts, Mr Watt soon produced some capital Engines; which were erected in Staffordshire, Shropshire, Warwickshire, and a small one at Stratford near London."[94]

We have hitherto seen Mr Watt as a man of genius, employed in improving a mechanism of infinite importance to his country and to mankind. When he had set the finishing hand to the invention, and it only remained for him to form those arrangements to secure to himself the reward of his labours:—his conduct as a man of business was as liberal and honourable with those who chose to avail themselves of his invention, as the mode he adopted to prevent disputes was satisfactory and ingenious.

The principle which was adopted by Mr Watt in granting licences to use his Engines, was to receive a *third part of the saving of coals* which was made by his Engines, when compared with an Atmospheric Engine doing the same work with coals of the same quality. A set of experiments were made to ascertain this point by persons of acknowledged integrity and skill: from knowing the depth of the mine, the size of the pumps, and number of strokes, which any engine made, either on the common or improved construction, they had only to estimate the saving of fuel in a certain number of strokes, and, according to the price at the stated intervals, they fixed their proportion. To tell the number of strokes, a small apparatus, consisting of a train of wheel-work, enclosed in a box, was fixed on the working-beam, so contrived, that each rise and fall of the beam moves one tooth of the small wheels, and this was shown by a small index. This mechanism was called the *counter*: it could only be opened by two different sets of keys, one of which was kept by the proprietors, and the other by Bolton and Watt, who had a traveller that went round to visit the different engines from time to time; and the counters being then opened and examined, from the number of strokes which had been made, the "patent third-part" was determined. This annual third could, however, be redeemed by a payment in one sum equal to ten years purchase. The counter was sometimes attached to what was called the spring beam floor, and a train of wheel-work was moved by the rise of the working-beam striking a small tappet or lever, which moved the wheel-work within.

In order to induce proprietors of mines, who were unable or unwilling to be at the expense of new Engines, Bolton and Watt at first took the old Atmospheric Engines in part-payment at a price far above their real value, and gave credit for the remainder until advantage should be experienced; and they even erected some Engines on several mines at their own expense, to be paid provided they answered the expectations, which they, as manufacturers, held out were obtained by their adoption. And it was subsequently stated that a sum not less than forty-seven thousand pounds had been expended in the speculation, by Bolton and Watt, before they began to receive any remuneration.

The mine of Chacewater was among the first at which they were introduced in Cornwall. Playfair has stated that here three of the largest-sized engines were erected; and it gives a high idea of the enormous expenses of the common engines, when we are told that for each of those machines the proprietors engaged to pay £800 annually, as a compromise for the patentee's "third part of the saving" made in coals by using Mr Watt's Engine instead of Newcomen's.

The expansive power of steam suggested by Mr Watt in 1769, and subsequently partially adopted to equalize the motion of the piston, was introduced as a means of saving steam in an engine at Soho manufactory in 1776, and in 1778 at Shadwell water-works, and afterwards particularly described in his specification of a patent in 1782. The parts of the *Expansive Engine* are in every respect the same with those we have already described; but the cylinders require to be made of greater dimensions.

If we now suppose the piston b, in the Twenty-fifth Figure, to be in the situation there represented, and the vacuum made under this piston by the usual means of opening a communication with the cylinder and the condenser; let the cock s be opened, and allow the steam from the boiler to enter and press down the piston until it reaches the middle of the cylinder, to about the position of a; and let the cock t be now turned, and allow no more steam to enter; the quantity which already has been admitted will press the piston to the bottom of the cylinder. It is also remarkable that this quantity of steam will do more work than if it had been allowed to flow into the piston to the limit of its stroke. Suppose, for instance, the boiler contains 100 cubic feet of steam, and that this quantity is to be expended in, raising a, certain quantity of water—say 100 cubic feet, when the communication is allowed to be open between the cylinder and boiler during its entire stroke; if this communication be shut off when the cylinder is only *half* filled with steam, the same quantity of 100 cubic feet of steam will raise 1,70 cubic feet of water. If the cylinder be only *one third filled with steam*, its effect will be greater than if it were filled one half; for the same quantity would raise 210 cubic feet of water; or, what is the same thing—

When the cylinder is quite full, its performance will be as	1·0
When 1/2 full, its performance is encreased as	1.7
When 1/3 full, its performance is encreased as	2.1
When 1/4 full, its performance is encreased as	2.4
When 1/5 full, its performance is encreased as	2.6
When 1/6 full, its performance is encreased as	2.8
When 1/7 full, its performance is encreased as	3.0
When 1/8 full, its performance is encreased as	3.2

upon the supposition that steam contracts and expands by variations, of pressure, in the same ratios that air would do. In practice it has been found, that under the natal pressure the encrease is not great after the steam is rarefied four times. It will be very obvious that this principle can he applied to both double and single engines; and that the common engines can be made to act expansively, or not, by the mere alteration of the tappets in the plug-frame.

In both these forms of the apparatus the steam acts not only to form the vacuum, but to depress the piston. But still, during the operation of the counterpoise it produces no effect; and where it was required to move machinery, this suspension of impulse was a great drawback omits utility. This, however, was not objected to its general merit, when used as a over of pumps; and the more so, as it was common to the Atmospheric Engine; but, before it could be considered, as a general first mover, this interval required to be much shortened, if it could not be altogether filled up.

We have seen that a contrivance was used at the Hartley colliery for the production of a continuous motion from a reciprocating one; but this did not obviate the inconvenience of the returning stroke; and it was left for Mr Watt to make this other step towards the perfection of the machine, and he accomplished it by a very slight extension of his first idea. He had introduced steam acting against a piston to press it downwards; he now formed a communication, between both sides of the piston and the boiler, and, also with the condenser; and made the steam act to press the piston *upwards* as well as *downwards*.

The mechanism was now, as far as the principle went, perfect; and it was freed, for the first time, from the enormous dead weight of counterpoises, which had hung on it from the first attempts of Newcomen; and the equally

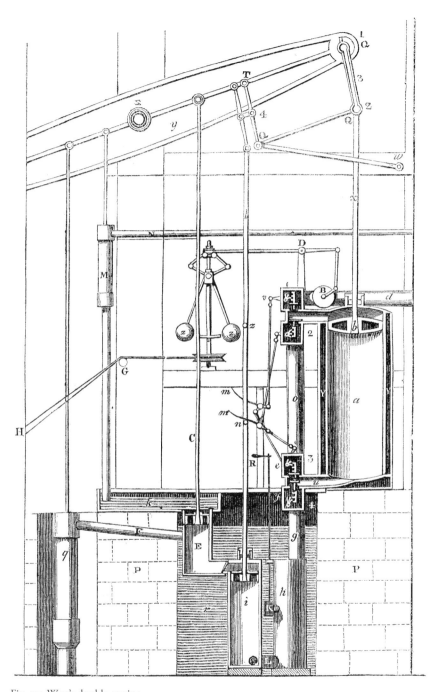

Fig. 27: Watt's double engine

enormous load which was used in the construction of the various parts, for the purpose of equalizing the motion.

The cylinder *a*, in Figure 27, is enclosed in a jacket or casing like the single engine, having a similar interval, which may be filled with steam or air. The piston *b* is attached to the lever-beam by the rod *x*. 1, 2, 3, 4, are the valves which admit steam to the cylinder, or open a communication between the upper and under sides of the piston, and the condenser. *g* is the pipe leading from the valves to the condenser. *m, m*, the levers or spanners, which are elevated or depressed by the tappets or pins *n, z*, in the plug-frame, and open or shut the valves to which they may be connected. *h* is the condenser; L, a pipe connecting it with air-pump *i*, and a second air-pump E. *c*, the piston-rod of this second pump, attached like the other, 1, to the lever-beam. *r*, a pipe from the cold water-pump *q*, to supply the reservoir in which the condenser and its pumps are placed. *k*, a trough or reservoir into which the water heated by the condensation of steam in the condenser, which is raised by the air-pump, is pumped back by M, into the boiler. G, a pulley; and H, an endless chain moving over it, also going round a pulley fixed on the upright axis of the *conical pendulum or governor z*. The other pulley, which is fixed to the axis of the fly-wheel, is not shown in this Figure, but its situation and connexion will be clearly seen in the following Figure. R, the handle of the lever which regulates the quantity of injection water admitted. P, P, the masonry or wall on which the cylinder and other parts of the machine are placed. *d*, a pipe conveying steam from the boiler to the cylinder. *n*, a cock or valve, called the throttle-valve or regulator, placed on the pipe conveying the steam from the boiler, and which is moved by the levers shown as supported at D, and connected with the conical pendulum. T, Q, Q, Q, W, are a system of levers called the parallel motion. *z* is the axis of the lever-beam *y*.

The motion at first is produced in this machine in the same manner as in the single engine by filling the condenser[95] and cylinder with steam, and then opening the injection-cock.

This process may be considered to have been gone through, and that the piston has arrived at the top of the cylinder. At this moment the tappets *n, n*, and levers *m, m*, open the valves 1 and 4, and shut 3 and 2. The steam from the boiler now acts on the upper side of the piston, while a vacuum is formed under it by valve 4 opening a communication between that. part of the cylinder beneath it and the condenser. The piston is therefore pressed by the elasticity of the steam to the bottom: when it has arrived at the lowest point, the tappets on the plug-frame, which also descend with the piston-rod, shut the valves 1 and 4, and open 3 and 2. The steam from the boiler, instead of flowing in at the top of the cylinder, is admitted at the bottom, and a communication is opened between the upper end of the cylinder and the condenser: a vacuum is then produced above the piston, and the elasticity of the steam (instead of the counterpoise in our last figure) forces it upwards. When it is elevated to the required height, the tappets again act on the spanners, and prevent the further flow of steam beneath the piston, and admits it at its upper end, opening at the same moment a communication between the lower end and condensing apparatus. The motion of the piston is then reversed, and this alternation may be continued indefinitely.

The mode of pumping out the water from the condenser being the same as that in the single engine, will be easily understood from an inspection of our figure. In order to show the four valves in section, a pipe placed in the same direction, and opposite to *o*, has been omitted in the engraving: it connects the top of the cylinder and the condenser together.

The power of the condensing Engine is easily known by ascertaining the temperature of the steam, which moves the piston; the area of the piston, and the temperature of the vapour which remains in the condenser. We know from experiment that steam of the temperature of 212° will balance the pressure of the atmosphere, or what is the same thing will force a piston into a vacuum with a force equal to about fourteen and three quarters pounds weight for every square inch of the area of the piston. The difference between the elasticity of the steam in the condenser, and that issuing from the boiler, will be the measure of the power of the engine. It is however found most expedient to raise the steam to a somewhat higher temperature than 212°, so as to produce a pressure between seventeen and eighteen pounds on each square inch of the piston; yet; in practice, from the imperfect vacuum which is made in the condenser; and after making allowance for the friction of the piston on the sides of the cylinder, and for the friction of the various parts of the intermediate machinery, this pressure of eighteen pounds on each square inch of the piston, cannot raise more water per inch than would weigh about eight pounds and a half; so that somewhat more than a half of the whole power of the steam is absorbed to give motion to the intermediate mechanism. The height to which this weight can be elevated depends on the length of the cylinder, and to raise the same weight to twice the height by the same piston requires double the quantity of steam. The double condensing Engine will also perform double the work of single Engine in about half the time, but it requires double the quantity of steam. So that in all cases, under the same circumstances, the work performed will be as the quantity of steam used.

When the impulse of the steam impelled the piston only in one direction, (downwards) its motion could be imparted to the beam by means of chains, as in the previous figure: but when the impulse was to be communicated upwards as well as downwards, some other contrivance for connecting the beam and piston became necessary; and one of the conditions of this contrivance must be to convert the motion in a curved path of the end of the lever-beam, into a rectilineal motion of the cylinder piston-rod. Mr Watt in his earlier engines used to form the end of the beam as a sector with *teeth*, which worked into a *rack* fixed on the end of the piston-rod: this allowed the rod to move perpendicularly upwards or downwards, but it was very inelegant in appearance, worked with a great noise, and was easily deranged, especially at the instant when the direction of the motion was changed.

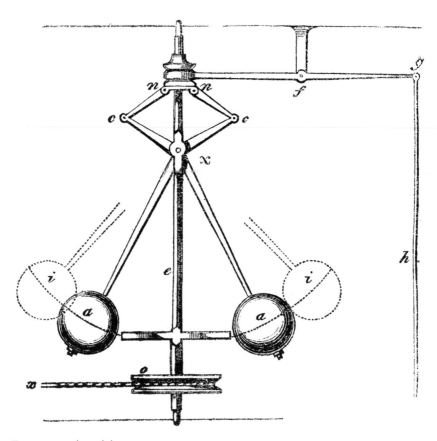

Fig. 28: conical pendulum

Even after the motion of the piston was equalized by shutting off the steam sooner or later from the cylinder, another source of irregularity was found to arise from the varying quantity of steam, which in different states of the fire under the boiler was admitted into the cylinder: several modes of adjustment occurred to Mr Watt. The one most generally employed, and probably as accurate as any, was by placing a valve in the pipe connecting the boiler and the cylinder, which could be made to encrease or diminish the steam-way. The next improvement was to make this valve, called a *throttle-valve*, a self-acting one, and to admit of its being so adjusted, that when the piston was-moving with too great a velocity, it would admit less steam into the cylinder, and so diminish its speed, and on the contrary admit a greater quantity when it was moving too slow.

A similar irregularity in the motion of corn-mills from the varying quantity of water or resistance, had early exercised the ingenuity of millers, to obtain some means by which its injurious effects could be obviated. One of the most usual modes was by means of a couple of heavy balls, attached by a jointed rod, which were made to revolve by being connected with the spindle or axis of the mill-stones. When the stones were moving at a great speed, the meal, by the rise of the stones, was too coarse; and on the contrary, when the motion was slow, the meal produced was small in quantity and too fine. The attached balls, which were called a *lift-tenter*, by their centrifugal force either raised or lowered a stage in which the arbour of the spindle revolved, and brought the mill-stones nearer, or removed them farther from each other, as they might be adjusted. This most ingenious regulator, was adopted by Mr Watt, and applied to regulate the opening and shutting of the throttle-valve of his improved engine. In Figure 27, z, z, are the two balls attached to an upright spindle by a joint, and which is connected to the throttle-valve by a series of small levers resting on a pivot, at D. It will be better understood by inspecting Figure 28. x, is an endless cord or chain going round a pulley, o, which is fixed to the vertical stem, e; retained in that position by moving in sockets at z and y. a, a, are two balls fixed to levers having a moveable joint at x: these levers are bent after their junction in the position shown in the figure, and are connected by another joint at c, c, to two short levers fixed by a joint at n, n, to a broad ring or strap of metal, moving freely up and down on the vertical spindle. On this ring or strap, formed as in the Figure, the ends (formed like a fork) of a lever f, are made to fit to it: this lever is suspended at f, and has another joint at g, by which it is connected to the rod h, which is attached to the axis of the throttle-valve. When the piston is moving at its usual speed, the balls, being connected by the endless rope x, will revolve and hang in the situation shown in the figure. But should the speed of the piston be much encreased, its motion will be communicated by the rope and pulley to the spindle *e,* and the balls will, by their centrifugal force, rise into the situation shown by the dotted lines *i,* i: this will depress the ends of their levers at *c, c,* and also the end of the horizontal lever which is attached to the moveable ring; and the rod *h*, will be raised, which closes in part the steam-way. Should the motion of the piston be too slow; instead of rising, the balls will fall, and have a contrary operation on the levers and throttle-valve; and diminish the quantity of steam admitted into the cylinder.

By these and other contrivances having made the reciprocating motion of his Engines very regular, Mr Watt early turned his attention to the important object of producing a continuous rotatory motion from a reciprocating one. We have seen that Stewart's scheme of employing ratchet-wheels for this purpose was useless, and Hulls was forgotten. "Among, the many schemes," says Mr Watt, "which passed through my mind, none appeared so likely to answer the purpose as the application of a crank in the manner of a common turning lathe (an invention of great merit, of which the humble inventor and even its era are unknown); but as the rotative motion is produced in that machine by the impulse given to the crank in the descent of the foot only, and is continued in its ascent by the momentum of the wheel, which here acts as a fly; and being unwilling to load my engine with a fly heavy enough to continue the motion during the ascent of the piston (and even were a counterweight employed to act, during that ascent, of a fly heavy enough to equalize the motion) I proposed to employ two engines, acting upon two cranks, fixed on the same axis at an angle of 120 degrees to one another, and a weight placed on the circumference of the fly at the same angle to each of the cranks, by which means a motion might be rendered nearly equal, and a very light fly would only be requisite."[96]

These ideas had very early occurred to Mr Watt; but his attention being chiefly directed to the making and erecting engines for raising water, he did not attempt to put them into practice until, about 1778 or 1779, when from the frequent breakages and irregularities of a mechanism similar to Fitzgerald's, for which Mr Wasbrough at Bristol had obtained a patent in this year, his attention was again drawn to the subject.

On trial it succeeded beyond his most sanguine expectations; but, having neglected to take out a patent, the invention was communicated, by a workman employed to make the model, to some of the people about Wasbrough's Engine, and a patent was taken out by, or rather in the name of Steed, for *the application of the crank* to Steam Engines. This fact was confessed by the workman; and also acknowledged by the engineer of Wasbrough who directed the works, he however excused his unprincipled appropriation, by urging that the same idea had occurred to himself prior to his hearing of Mr Watt's; and he further averred that he had made a model of it before that time: "this," with great candour says Mr Watt, "might be a fact, as the application of it to a *single crank* was sufficiently obvious; and in these circumstances I thought it better to accomplish the same end by other means, than to enter into litigation, and, by demolishing the patent, to lay the matter open to everybody."[97]

In order to obtain a rotative motion from a rectilineal one by some other means than the crank, Mr Watt introduced what is now called the *Sun and Planet Wheels*: it is also a cardioid motion, and has several advantages over the crank, where a more rapid velocity is to be given to the fly; the fly being made to revolve with double the speed as when a crank is employed. It is not, however, either so simple, its construction makes it more expensive, and it is easily put out of order, and has now given place to the crank. *x,* in Figure 29, is an arm attached to the lever (or working) beam; and *a,* the axis which imparts motion to the machinery. The wheel *b,* is fixed to *x,* and the wheel *a,* on the axis of *z*; and both wheels are fixed as nearly as possible in the same vertical plane. At the rise and fall of the working-beam, the wheel *b,* is carried round the circumference of the wheel *a,* and at the instant when the motion of the lever is reversed, the continuous rotatory movement is produced by the momentum of the fly-wheel.

In many cases it was of importance to ascertain the degree of exhaustion of the condenser, or rather the elasticity of the vapour remaining in it; this Mr Watt accomplished by adapting a barometer tube filled with mercury (open at both ends) to the condenser, and about thirty inches long. Its action is nearly the same with the common barometer; were the vacuum in the condenser perfect, the mercury would be raised in the tube to the barometric height of the atmosphere, but as this is never attained in common instances in the condensers of Steam Engines, the mercury in the tube is raised to a height corresponding to the degree of its elasticity, or exhaustion. A similar tube, called the Steam Gauge, was fixed on the boiler, which indicated by the rise and fall of the mercury, the elasticity of the steam; that on the condenser showing the force with which the atmosphere pressed on the mercury to *enter the condenser*; while the rise of the mercury in the tube placed on the boiler shows the force with which the steam presses it to *enter the atmosphere*, at all temperatures above 212 degrees: below this, its action is the same as the condenser gauge.

"It once happened that the valve of the pump-bucket, breaking the engine suddenly lost its load or resistance, which occasioned the piston to descend

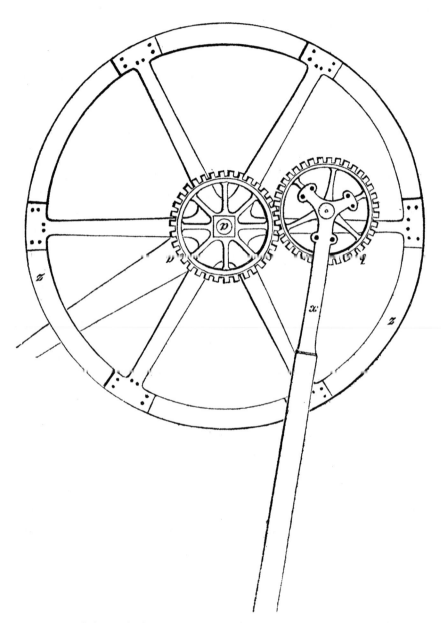

Fig. 29: sun and planet wheels

and strike on the spring beams for two or three successive strokes with such violence as to break one of the beams, and at last the piston striking the bottom of the cylinder, the momentum of the beam forced down upon the rod so violently as to bend the great piston-rod quite crooked. To prevent similar accidents, a smaller steam pipe was added at the side of the vertical steam pipe communicating with the passage into the bottom of the cylinder. This pipe is kept closed by a valve; but, if the engine descends so low as to strike on the spring beam, a catch pin on the beam strikes a small lever, and by a wire of communication opens the valve and lets the steam into the lower part of the cylinder beneath the piston, and this destroys the vacuum, so as to prevent the further descent of the piston.[98]

The scheme of equalizing the motion of the piston by a fly, suggested by Fitzgerald, and mentioned by Mr Watt in 1769, in the letter to Dr Small which we have quoted, appears to have been first used by Mr Wasbrough, and it was applied in combination with the crank in his works at Bristol for turning lathes, and also at Mr Taylor's works at Southampton; and he likewise applied it to some corn mills.[99]

In the drawing produced by Mr Watt before the House of Commons in 1774, he shewed a method of applying the steam to press on both sides of the piston. This was brought forward by a Dr Falck in 1779 as a new idea. The Doctor published a pamphlet enumerating the advantages of this kind of engine, stating it to be equal to perform double the work of the usual lever Engines of the same size; but he does not appear to have proved the assertion by constructing a model.

The only novelty in the Doctor's scheme was the way he produced this double action. He proposed using *two cylinders* fitted with pistons, connected with the same boiler. Steam was introduced under both pistons, and a vacuum was formed by its condensation, by the same methods as those practised in the Atmospheric Engines. Only, that while the steam was admitted from the boiler into one cylinder, it was prevented from flowing into the other by a regulator. The piston-rods were kept (by means of a wheel fixed into an arbour) in a continual ascending and descending motion in the same manner as the rods of a common air-pump, by which they move a common axle, having another wheel affixed to it; this wheel moves the pump rods, in the same alternate directions as the piston rods, by which the pistons of the pumps are kept in constant action. "We have seen both the Atmospheric and single Engine of Mr Watt working in this arrangement, but his double Engine is much preferable."[100]

About this time Mr Watt's condensing Engine was introduced into France. In 1779 a M. Perrier, who was associated with his brother in the manufacture of machines at Paris, was commissioned by a company then recently established to supply that city with water, to proceed to England and to endeavour to procure a Steam Engine manufactured on the best principle; those which were constructed in France being exceedingly defective in arrangement and workmanship. M. Perrier ultimately succeeded in purchasing from Bolton and Watt an engine in which were combined their improvements. By the sanction of the English Government, it was sent to France, and erected by MM. Perrier at Chaillot, near Paris. The French engineer Prony, in his elaborate work on the Steam Engine, describes this identical engine; is minute even to prolixity in his enumeration of the peculiarities of its construction and its economical advantages; yet with a detestable illiberality and utter disregard to everything like candour or truth, he attributes all the merit of the improvements in the Chaillot Engine—to his friend Perrier, the person who merely put together the pieces he had brought from Soho, without once alluding to Mr Watt's labours, or once mentioning his name in his quarto book, as being the person who constructed it,[101] or even as having made any improvement in steam mechanism! This attempt to award to his countryman the honour due to another is entitled to still more unqualified reprobation, from M. Prony's personal knowledge of the whole transaction; for we believe he was the person under whose immediate direction Perrier was sent to England, and who was in correspondence with that gentleman when he was residing at Birmingham, during the execution of the Chaillot Engine; and who superintended Perrier when he put the Engine together in France.[102]

The Engine invented by Mr John Hornblower presents no novelty in its principle or even in the arrangement, combining Mr Watt's expansion Engine with the two cylinders of Dr Falck. The steam, after it was admitted to elevate or depress the piston in the first cylinder, was then admitted to perform the same action in the second cylinder, which in dimension considerably exceeded the first cylinder, and thus was allowed to operate by its expansion to raise the piston. Figure 30 will give a general idea of this contrivance, without showing any of the smaller parts, which, as being similar to those of common engines, and not to confuse the Figure, are omitted. A, the small cylinder, having a piston, *a*, attached by a rod to the lever-beam; this cylinder communicates with the boiler by the pipe *c*. Other pipes and cocks, X, Y, (Hornblower arranged his parts more compactly, but, in order to explain this machine in one Figure, we arrange them somewhat differently,) are attached to each cylinder, and open a communication with both sides of their pistons. E, is a pipe with a stop cock, which opens a communication between the bottom of the small cylinder A, and the top of the large cylinder B. F, is a pipe and cock leading

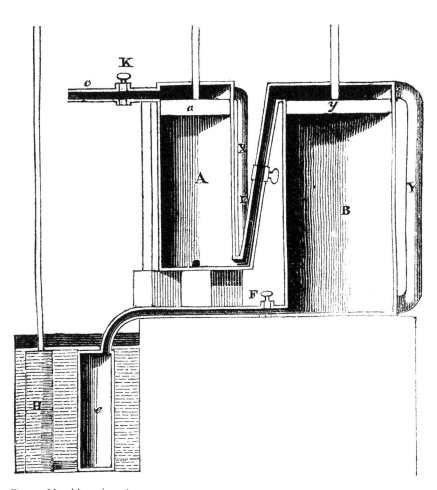

Fig. 30: Hornblower's engine

to the condenser G, its pump H, placed as usual in cold water; these are the same as those in Mr Watt's Engines. Steam comes from the boiler by the pipe *c*, and its flow may be prevented by the cock K. We will now suppose the cock K, to allow the passage of steam from the boiler, and the cocks E, X, Y, to be all open, which will allow the steam to fill both cylinders. The cocks X and Y must now be closed, E and K remaining open.

By turning cock F, a communication is opened between the under side of the piston Y, and the condenser, which forms a vacuum in B. The steam pressing on the upper side of *a*, in A, and the communication between the cylinder A under its piston *a* being open, the steam in A, from its expansive power, will press Y, downwards in B. This decreases the resistance on the under side of the piston *a*, which is also carried downwards by the pressure of the steam flowing through K from the boiler; and the two pistons descend at the same time, carrying the beam along with them. When they reach the bottom of the cylinders, the cock F, shuts off the communication with the condenser, and the cock E, with the top and bottom of the two cylinders; the cocks in Y, and X, are now opened, which allows the steam in each a free communication between the upper and under sides of their pistons, or between that of the last cylinder and condenser; and the counterpoise at the other end of the lever-beam raises the pistons to the top of their respective cylinders; or the pipe Y, may form a communication between the top and bottom of piston *y*, and the condenser. X and Y are now shut, and E, F, and K, are opened, and the operation is repeated.—The mode of admitting the steam to allow of the action of the counterpoise, the shutting out the atmosphere, the condenser, and the air-pumps connected with it, are taken from Mr Watt's Engines. It is unnecessary to enter further into the detail of this engine, which was not found in practice to be at all equal to the common form of the condensing Engine. In fact, the apparatus described by Hornblower is in every part the same as Mr Watt's.[103] And it must always be subject of regret that this ingenious man should have wasted the best part of his life, and ruined his fortune, in a series of selfish attempts to copy Mr Watt's inventions without coming within the letter of the patent.

Mr Watt, who had previously constructed engines on the principle of the double impulse, finding inserted himself beset with a host of plagiaries and pirates, inserted a description of it in a patent which he obtained in March 1782. He also described a compound Engine, or method of connecting the cylinders and condensers of two or more separate engines, in which the steam, after it had pressed on the piston of the first engine, was made to act by its expansion to press on the piston of the second engine, and thus to derive additional power, which might be employed to act either conjointly or alternately with the piston in the first cylinder. The reciprocating rotative engine, and the rotative engine[104] also described in this patent, he never carried into execution. The mechanism for transmitting the double impulse of his new engines, used for a few years, was given up for less objectionable methods.

The application of the Engine to the moving of Mill machinery gave birth to another contrivance. In the patent of 1782, the double impulse was communicated to the working beam by the intervention of a sector placed on the end of the pump-rod, working into a sector placed on the end of the working-beam; but the motion was rough and jerking, and above all noisy; and the racks and sectors were very subject to wear. It occurred to Mr Watt that if some mechanise could be devised, moving on centres which would keep the piston-rods perpendicular, both in *pulling* and *pushing*, a smoother motion would be obtained. This problem was solved by the invention of the beautiful mechanical combination called the *parallel motion*. It is shown as attached to Figure 27, and was first used in the engines erected at the Albion Mills near London.

This document also contained a description of several contrivances for regulating the descent of the piston: one of them was by using two cylinders open at top and bottom, and also furnished with two pistons suspended from the opposite ends of a lever moving on a fulcrum; the cylinders are connected by a trough at the top, and are each filled with water. When the pistons are in equilibrio, and the connecting lever horizontal, the height of the water above each is equal; but when either end is depressed, the water raised by the opposite piston flows through the trough into the other cylinder, and this difference of pressure being added to the resistances of the steam piston in its ascent or descent, equalizes with tolerable precision its otherwise accelerated movement:—levers acting unequally on each other, and chains acting on a spiral in the manner of a fuse, and a large weight attached to the working-beam, having its centre of gravity considerably above its centre of motion, are also proposed as modes of equalizing the rise and fall of the piston. They are all ingenious; but except the water cylinders, they are not much adapted to practice, and we believe they were not employed by Mr Watt himself in any of his engines.

Mr Watt's next improvement was made in the construction of the fireplace; and his invention is the type of a great proportion of those contrivances for consuming the smoke, which have been proposed by numerous, recent projectors, on the principle of the air which is to supply part of the combustion, being made to traverse the surface of the ignited fuel before it could come into contact with the boiler, or ascend the chimney. The whole arrangement of this fire-place is very good. A similar mode of supplying the fuel by a kind of hopper, and also the two side openings by which air was admitted, but on a small scale, was practised by the alchemists: the furnace in their hands was called the *Athanor*. Glauber used a similar one, and describes it in his writings. Mr Watt also proposed using *two* fires, one placed beyond the other; on the first grate was laid coals; and the smoke from this fire was made to pass over the surface of the fire made on the other grate with charcoal or coke; so that whatever combustible gas escaped from the fuel on the first grate was ignited by the burning cokes placed on the second.

The first practical application, on a large scale, of the Steam Engine to propelling vessels, was made about this time by a French nobleman.[105] It appears that a Marquis de Joufroy (Geoffroi?) in 1781 made some experiments on a great scale, on the river Saone at Lyons, with a boat stated[106] to have been 140 feet long. We are ignorant of the details and arrangement of the mechanism of this vessel, and equally so of the circumstances which occasioned the scheme to be abandoned. The same idea, it would seem, a little after this period occurred to a Mr Miller, of Dalswinton near Edinburgh, an amateur mechanic, of very considerable scientific acquirements, and of great ingenuity, who published a description of what he called a *triple boat*, and presented copies of his pamphlet to the different Sovereigns of Europe. The vessel which he proposed to propel by a Steam Engine was a *double boat*, having *one wheel*[107] in the middle. It was repeatedly tried on the Forth and Clyde canal, and the experiment, it would appear, was successful.[108] The date of the trial is not given, but it was probably made after the publication of Mr Miller's book in 1787, as be there speaks of the scheme *as likely to answer*.[109] That this gentleman was the *inventor of the Steam Boat* in the strict sense of the word, "I will not," says Dr Brewster, "venture to affirm; but I have no hesitation in stating it as my decided opinion, that he is more entitled to this distinction than any other individual who has yet been named!"

M. Bettancourt, whose name is well known among mechanics as the author of some experiments and formula: on the elastic force of steam, was employed in 1787 and 1788 by the Court of Spain to obtain information and collect models of machines, which it might be expedient to introduce into the Spanish-American mines. When in England, he took occasion to visit the Albion Mills, in which Mr Watt had erected one of his double impulse engines. Here M. Bettancourt observed that instead of chains, the parallel motion was introduced, from which he inferred that the piston was impelled by the steam, both upwards and downwards; but the interior mechanism he

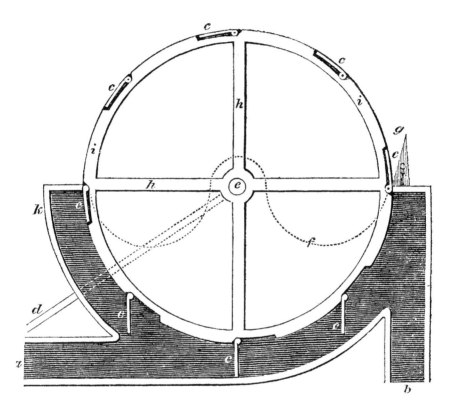

Fig. 31: Cooke's rotatory engine

had no means of inspecting. At his return to Paris, he constructed a model having the external appearance of Mr Watt's Engine; but the steam valves, and the mode of connecting the cylinder and the boiler, were his own invention: these are very imperfect. An engine which is fully described by Prony, was erected by M. Perrier near Paris, from Bettancourt's model: neither its merit, nor date, entitles it however to any notice in the history of the Steam Engine, but for the claim which Prony makes for Bettancourt, as being a second-hand *inventor* of the mechanism of the double Engine.

Mr Cooke presented a description of a rotative Engine to the Royal Irish Academy in 1787, which is shown in Figure 31. On the circumference of a wheel eight vanes or flaps are attached by joints, which are formed to open somewhat more than half of their circumference. During the revolution of the wheel, the valves which are on the lower half of its circumference, hang in a vertical direction by their own gravity. $c, c, c,$ are the valves or flaps; $b,$ is the tube which admits steam from the boiler; $a,$ a tube leading to the condenser. $k, k, f,$ is the case in which the wheel $h, h,$ is enclosed as high as the dotted line: this case is to be steam-tight. The wheel being supposed in the situation in the Figure, the valves prevent any communication between the boiler and condenser. Steam is now admitted at $b,$ and pressing on $c, c,$ forces them forward in its passage to the condenser, and produces a rotatory movement. The condenser is worked by a crank in the axis, and a rod d is extended from it, which keeps a constant vacuum in that half of the steam case:—"by this means a power is added to the steam equal to the weight of the atmosphere so that when the force of the steam is only equal to the pressure of the atmosphere, and the valves are six inches square, the wheel will be forced round by a power equal to 531¼ pounds placed on its circumference,"—if, as Mr Cooke ought to have added, his scheme was practicable with such imperfect mechanism. $g,$ is a piece of wood which "shuts the valves as they approach, and delivers them closed into the steam vessel."[110]

The contrivance which has been described for consuming smoke in furnaces was about 1790, introduced into France by Mr Watt. "We should have been ignorant," say MM. Morveau and de Prony, (in their report on the method employed in the French Mint of consuming the smoke of its engine,) "of Mr Watt's mode of consuming smoke, if he had not adapted it to the apparatus of a Steam Engine at Nantes, the several parts of which were got up in his workshops, and which was erected at Nantes in 1790, under the direction of our fellow member, who had discussed and arranged the plans with Mr Watt himself."[111] In this year also the first rotative condensing engine, used as a first mover of cotton machinery, was erected in the town of Manchester.

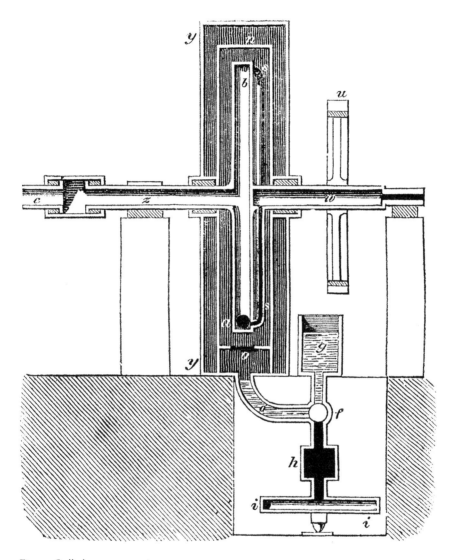

Fig. 32: Sadler's rotatory engine

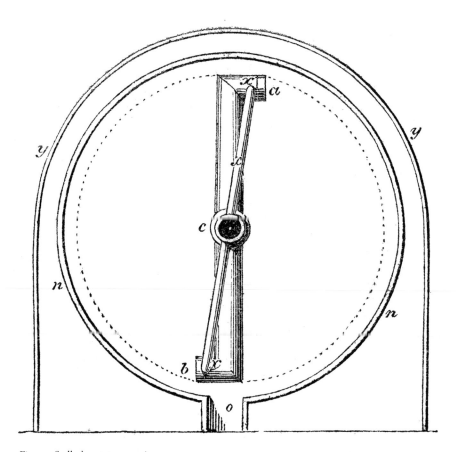

Fig. 33: Sadler's rotatory engine

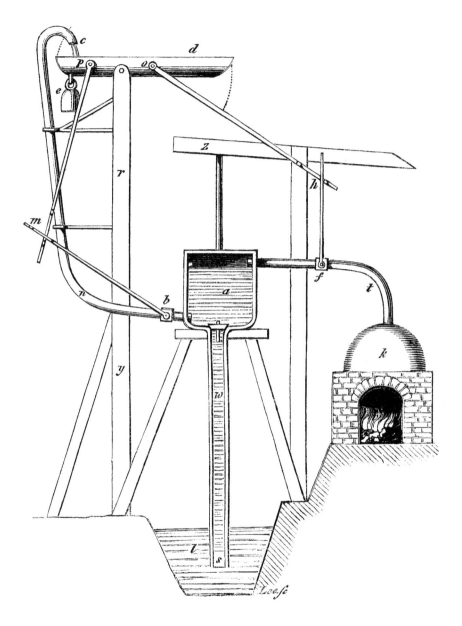

Fig. 34: Francois' engine

In Mr Sadler's invention,[112] (Figures 32 and 33,) the steam generated in the boiler is conveyed through the pipe *c*, into the spindle or axis of the rotative cylinder *a, b,* which is made steam-tight by working in a stuffing-box. The steam passes along the arms of the revolving cylinder, nearly to its ends, where it meets with a jet of cold water, introduced from the hollow axis *w*, by the small pipes *s, s*; this condensing water falls from the revolving cylinder into the bottom of case *n*, at *o*; whence it is conveyed through a pipe *o*, and cock *f*, into *h*, and is discharged by openings made in the ends or sides of another cylinder *i, i,* moveable in a horizontal direction, giving it a rotatory movement in the same manlier as Barker's mill. The jet of cold water from the pipes, *s, s,* having condensed the steam, produces a reaction, and the cylinder *a, b,* acquires a rotative movement. The case *n*, is steam-tight, and the outer case, *y,* serves the same purpose with the jacket in the reciprocating engines. Another mode of action is suggested by Mr Sadler to be had by filling the case (in which the arms revolve) with steam, which would cause them to revolve by the pressure it would produce in being condensed in entering the arms. Figure 33 is a side view of this steam wheel: *x, x, x,* being the small pipes for conveying the injection from the hollow pipe *c,* to the end of the arms *a, b*. *n,* is the enclosing case, and *y,* the jacket; leaving an interval between them which may be filled with any slow conducting substance. *o,* is the pipe leading to the revolving arms which discharge the hot water. This Figure may be considered as an illustration of the engine already noticed as having been tried by Mr Watt, but with which, it is probable, the ingenious inventor was unacquainted.

M. Francois, professor of philosophy at Lausanne, having been consulted by some members of the government, respecting the draining of a considerable extent of marshy land, between the Lakes of Neuchatel, Bienne, and Moral, and from some local peculiarities having given up the idea of accomplishing this object by the use of windmills, proposed using a Steam Engine on the plan of Savery's.

The machine (Figure 34) is composed of the pipe *s*; *w,* the lower end of which is inserted into the water, and the upper end enters the receiver *a*; a pipe *n, c,* proceeds from the receiver as high as it is required to elevate the water, and has a cock or valve at *b*. Another pipe *t,* having a stopcock at *f,* conducts steam from the boiler *k*, into the receiver *a*. *d,* is a bucket turning on an axis: to this bucket two levers are fixed, having joints at *p,* and *o*. *z,* a trough, which conducts the water, which may be raised to the required level from the ditch or excavation *l*.

The novelty in this engine consists in the simple means by which it is made to be self-acting. When the steam, flowing from the boiler into the receiver, forces the water which it may contain up the pipe *n*, and is emptied into the swinging bucket *d;* this bucket being filled turns upon its axis, and empties itself into the trough *z*, and is afterwards brought into its horizontal position by the counterpoise *e*. However, as the bucket turned down, it depressed the lever *o*, and raised the lever *p*; the lever *o*, acted on *h*, to shut the cock *f,* and the lever *p*, acted on *m*, to shut the cock *b*: the further passage of steam from the boiler into the receiver was shut, and the water in the eduction-pipe prevented from flowing into the receiver. A valve at *z,* in the trough, was also at this moment raised, which permitted a small quantity of the cold water to fall into *a*, and condense the steam it contained. The pressure of the atmosphere then forced the water into the receiver, and the counterpoise restoring the swinging bucket to its former position, the cocks *f,* and *b*, are opened; "the steam escapes from the boiler, and the water is again driven out, and this takes place five or six times in a minute."[113]

Kempel's rotatory[114] Engine, described by Lansdorf, differs nothing in its principle from Hero's. A boiler *a*, in Figure 35, is fitted with a cock *d*, and vertical pipe, to which is attached a horizontal arm, *c, x, b*, moving round the vertical pipe at the joint *x*. Steam, being admitted from the boiler through the cock *d*, it is allowed to escape into the atmosphere from two small holes

Fig. 35: Kempel's rotatory engine

made in the side of the horizontal pipe, one on each side of the joint, and by its re-action the arms were carried round continually. At the ends of the horizontal arm, two circular pieces were fastened, which served the purpose of a fly-wheel, and the rotative motion was communicated to other machinery by a chain or cord connected with the upright pin *x*.

The earliest mode of condensation we have described as being produced by an allusion of cold water on the outside of the cylinder; and it was also attempted by Mr Watt by immersing his condenser in cold water: both schemes were found to be ineffectual. The Reverend Mr Cartwright,[115] in reviving the scheme of condensing by contact, suggested the advantage of an engine in which this could be accomplished, as saving, in many cases, nearly the whole expense of the fuel. In distilleries, for instance, the alcoholic vapour might be introduced under the piston, to raise and depress it by its elasticity, and be condensed without mixing with cold water, in the same way as in the worm of a still; or, should this not be required, the boiler could be filled with alcohol; and from the lower temperature required for its conversion into vapour, with an elasticity equal to the pressure of atmosphere, Mr Cartwright calculated at least a saving of one half of the fuel could be made. The details of the engine he proposed to fulfil these conditions were constructed with uncommon ingenuity; and the whole apparatus may be considered more simple and efficient than any other combination which had been proposed of the parts of the condensing engine.

In Figure 36, the piston *b*, moving in the cylinder *a*, has its rod *t* prolonged downwards; another piston *d*, is attached to it, moving in the cylinder *e*, and which may be also considered as a prolongation of the steam cylinder. The steam cylinder is attached by the pipe *g*, to the condenser, placed in cold water, formed of two concentric circular vessels, between which the steam is admitted in a thin sheet, and is condensed by coming into contact with the cold sides of the condensing vessel. The water of condensation falls into the pipe *e*. To the bottom of the cylinder *i*, a pipe in is carried into a box *n*, having a float-ball *o*, which opens and shuts the valve *p*, communicating with the atmosphere; a pipe *q*, is also fitted to the box. There is a valve placed at *i*, opening into the cylinder *e*; another at *n*, also opening upwards. The pipe *s* conveys steam from the boiler into the cylinder, which may be shut by the fall of the clack *r*. K, is a valve made in the piston *b*.

In the Figure the piston *b* is shown as descending by the elasticity of the steam flowing from the boiler through *s*; the piston *d*, being attached to the same rod, is also descending, When the piston *b* reaches the bottom of the cylinder *a*, the tail or spindle of the valve *k*, being pressed upwards, opens

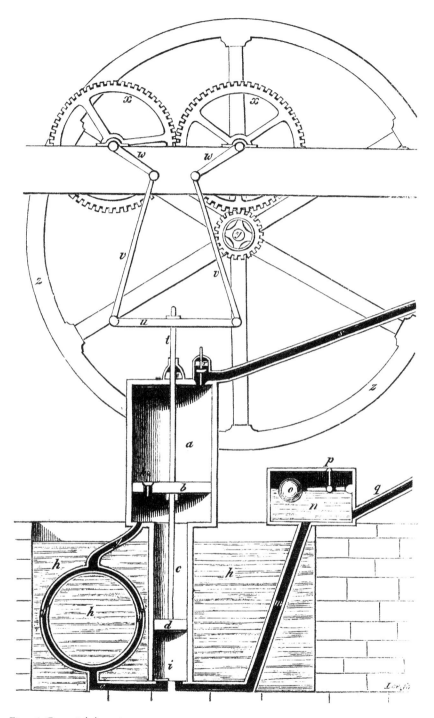

Fig. 36: Cartwright's engine

the valve, and forms a communication between the upper side of the piston and the condenser; at the same moment the valve *r* is pressed into its seat by the descent of the cross arm on the piston, which prevents the further admission of steam from the boiler; this allows the piston to be drawn up to the top of the cylinder, by the momentum of the fly-wheel *z*, in a non-resisting medium. The piston *d* is also drawn up to the top of *c*, and the valve *i* is raised by the condensed water and air which have accumulated in *e*, and in the condenser *g*. At the moment when the piston has reached the top of the cylinder, the valve *k* is pressed into its place by the pin or tail striking the cylinder cover; and at the same time the piston *b* striking the tail of the valve *r*, opens it; a communication is again established between the boiler and piston, and it is forced to the bottom as before. By the descent of the piston *d*, the water and air which were under it in the cylinder *c*, being prevented from returning into the condensing cylinder by the valve under *i*, are driven up the pipe *m*, into the box *n*, and are conveyed into the boiler again through the pipe *q*. The air rises above the water in *a*, and, when by its accumulation its pressure is increased, it presses the float *o* downwards; this opens the valve *p*, and allows it to escape into the atmosphere.

The machine, from its refined simplicity, appears excellently adapted as a first mover on a small scale. It has never, however, had a fair trial; the objections which were urged against the condensing vessels at the time of their invention, have always appeared to us more specious than solid. To the great merit in the arrangement and simplification of parts shown in this engine, must be added one of immense importance to engines on every construction—*the metallic piston*. Mr Cartwright constructed his of two plates, between which were placed detached pieces of metal, instead of the usual packing; these pieces were acted on by a spring, so as always to be kept equally tight, whatever might be the wear of the piston. The manner of connecting the piston-rod and procuring a rotatory motion is a beautiful specimen of mechanical invention.

In this patent Mr Cartwright also proposed a modification of the rotatory Engine described by Mr Watt in 1782. It is shown in Figure 37. The axis D has any unequal number. of pistons, H, H, H, (not less than three) upon it, so disposed that no two shall be opposite. In the cylinder in which the axis revolves, are two valves; at opposite sides of the cylinder; on the side of each valve is a pipe for the admission of steam, and another communicating with the condenser. It is obvious that the number of pistons and valves not corresponding, some one or other of them will always be in action on one side or the other of the cylinder. The engine conse-

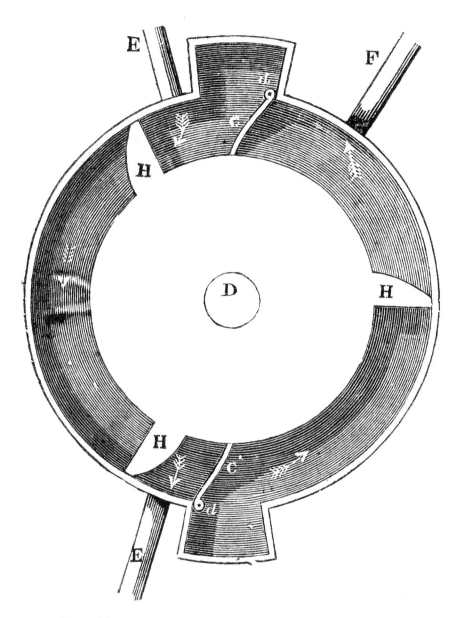

Fig. 37: Cartwright's rotatory engine

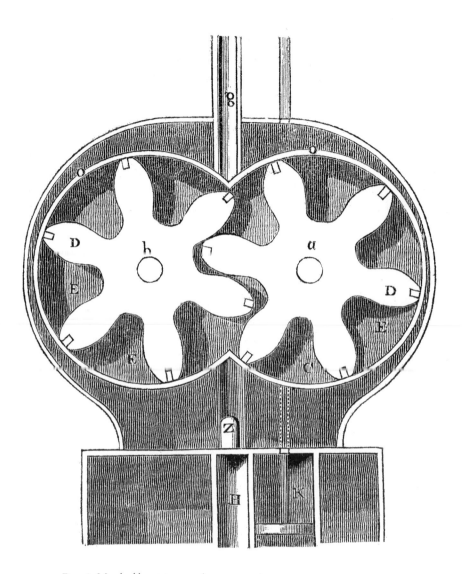

Fig. 38: Murdock's rotatory engine

quently will have no tendency to stop, and therefore a fly-wheel will not be absolutely necessary, especially a very heavy one.[116]

Mr W. Symington, who was employed about Mr Miller's Steam Boat, constructed in 1801 a vessel exhibiting considerable ingenuity in the arrangement and construction of the steam machinery. He placed the cylinder in a position nearly horizontal; the piston was supported by friction-wheels, and, as the lever-beam was dismissed, he communicated the motion to the paddle-wheels by a rod and crank attached to the piston. The paddle-wheel (as in Mr Miller's boat) was placed in the centre of the vessel; he also attached stampers to the bow of his boat for the purpose of breaking the ice on the canals. This apparatus moved the vessel at the speed of only about two and a half miles an hour, but was laid aside in consequence of the injury, even with this small velocity, which was apprehended to the banks, from the eddies raised, by the action of the wheels in the water.[117] Mr Buchanan was unable to ascertain whether he ever tried his boat on any river.[118]

Next in importance to Mr Watts improvements on the Engine, may be reckoned Mr Matthew Murray's, of Leeds, on the self-acting apparatus attached to the boiler. In 1799, this gentleman connected the damper of the chimney with a small piston moving in a cylinder, which rose and fell with the increased or diminished elasticity of the steam in the boiler, and opened and shut the damper attached to it, and by this means regulated the intensity of the fire under the boiler—an invention of great practical use, and among the few which are still used on all well-constructed Steam Boilers; he also revived the old sliding valves with great improvements; gave a new arrangement to some of the other parts, and greatly improved the air-pump, besides introducing several improvements in the details of the many beautiful engines which were constructed in his great manufactory at Leeds. Mr Murray, we believe, also carried into practice the scheme of placing the piston in a horizontal position in the common condensing engines. About the same time Mr W. Murdock formed the upper and lower valves with one spindle; and by forming the connecting tube hollow, he made it serve the office of the eduction-connecting pipe between the top and bottom of the steam cylinder; and by this ingenious contrivance freed the engine from a pair of valves. The same patent also included a description of a rotative Engine, which is shown in Figure 38. *a,* and *b,* are two wheels working into each other, and turn in a box, to the inner circumference of which the ends of the wheels are made to move round steam-tight by packings or stuffings, as shown in our Figure. The axis of one or both wheels pass through the sides of the box, and are made steam-tight in the usual manner. Steam being admitted into this box by the

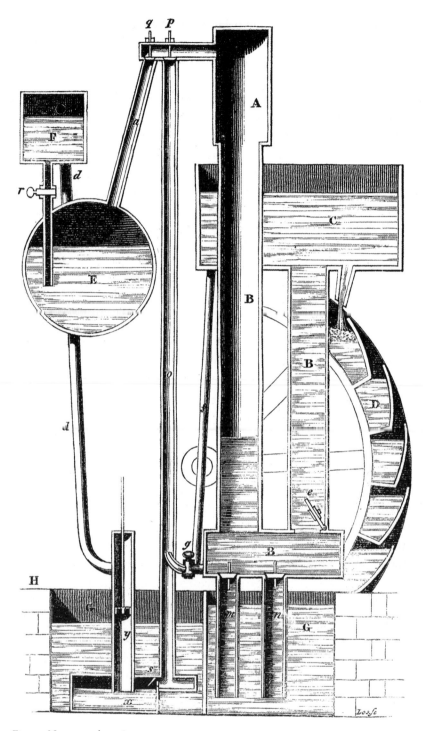

Fig. 39: Nuncarrow's engine

pipe or channel *g*, and a vacuum being formed by the condenser *h,* the steam acts upon the teeth of the wheels, and causing them to turn round in contrary directions, produces a rotatory motion, which may be communicated from their axes to other machines[119]—Mr Bramah, in 1801, made a considerable improvement on the fourway-cock, which had been used in low-pressure engines, by making it revolve continuously instead of the usual alternate revolution. The improvement was important in two respects: its wear was less, and of course its action capable of greater precision.[120]

An improvement on Savery's Engine by condensing the steam in a separate pipe or vessel from the body of the pump, was described by Mr John Nunearrow in 1799.[121] The pipes *m, n,* (Figure 39) are inserted in a cistern; *a*, a steam pipe through which the steam is conveyed into A, B, the receiver; *o*, the condenser; *n*, the boiler; *r*, a reservoir for supplying the boiler with water. G, the injection-pipe, and *g*, its cock; *y,* a pump by which water is raised from the cistern through pipe *d,* into E. C, the reservoir from which the water, which is raised through the pipe B, falls into the buckets of the water-wheel D, and gives it a rotatory motion.

The steam is admitted through pipe *a*, into A, B, and pressing on the water which it contains, the water is raised through A, B, and B, into the cistern C. The valve *q* being now shut, and *p* suddenly opened, the steam rushes down the pipe *o*, and, meeting with a jet of cold water from the cock *g*, is condensed. A vacuum being by this means made in the receiver, the water is forced up by the air's pressure, and again fills it, when the valve *q* is opened, and the elasticity of the steam again raises the water into the reservoir *c,* through B—the condensed steam is allowed to fall into the well or box *x,* and it is pumped into the boiler. The other parts of this machine will be easily understood from inspection. In the engraving, the cock *q* is shown as shut; it should have been represented as open.

The project of Dr Robison in 1759, to apply the Steam Engine to move carriages, was accomplished in 1802 by Messrs Vivian and Trevithick; and in carrying the speculation into practice, they produced a Steam Engine of pre-eminent merit and ingenuity. The principle of moving a piston by the elasticity of steam against the pressure only of the atmosphere, was described by Leupold, but in his apparatus it acted only on one side of the piston. In Trevithick and Vivian's engine, the piston is not only raised but depressed by the elasticity of the steam; and although other modes of admitting the vapour may be employed, they gave the preference to Leupold's[122] fourway-cock. Mr Watt too had an idea of impelling the piston on both sides by steam against the atmosphere, and this was included in one of his patents; but it is equally certain that he never carried it into practice. Although, therefore, the ingenious Cornwall engineers do not claim the merit of inventing the high-pressure engine, yet for the beautiful simplicity, great portability, admirable selection of forms, and ingenuity in the arrangement, they are entitled to all praise; and their apparatus may be classed with Savery's, Newcomen's, and Watt's, as forming an era in the history of the mechanism.

The immediate object of Messrs Trevithick and Vivian was the production of a portable mechanism to be applied to carriages, and so far they succeeded; but their engines are equally well adapted to every use to which those of Mr Watt are at present exclusively applied, and with the most ordinary precaution they can be made as safe from accident as those called in contra-distinction low-pressure; indeed, notwithstanding their recent invention, the consequent want of experience in their management, and the operation of vulgar and absurd prejudices against their safety, they may now be considered the only formidable rivals to the condensing Engine.[123]

The experiment of moving carriages was successful, and in 1804 one of these locomotive Engines was in use at a mine at Merthyr Tydvil, in South Wales, and drew as many carriages as contained about ten tons and a half of iron, travelling at the rate of five miles and a half an hour, for a distance of nine miles, without any additional water being required during the time of its journey—its cylinder was eight inches diameter, and the piston had a four feet stroke. The same engine, employed to raise water, worked a pump of eighteen inches and a half diameter and four feet and a half stroke, raising it twenty-eight feet high, and made eighteen strokes in a minute; it used eighty pounds weight of coal an hour, and in the same time it raised 15,875,160 pounds weight of water one foot high; the pressure being sixty-five pounds on each square inch of the piston.

We have said that Trevithick's Engine is the most compact, it is also the simplest in its operation. Steam is admitted from the boiler under a piston moving in a cylinder, which impels it upward; when it has reached this limit, the communication between the piston and under-side of the piston is shut off; and the steam which has raised the piston is allowed to escape into the atmosphere; a passage is opened between the boiler and the upper side of the piston, which is then pressed downwards; and the steam is again allowed to escape into the atmosphere. From this it is obvious that its power will be equal to the difference between the pressure of the atmosphere and the elasticity of the steam. Figure 40 will explain the operation of Trevithick and Vivian's apparatus; the parts are, however, to be considered, as placed to explain their operation in the clearest manner, rather than in the situation in which they

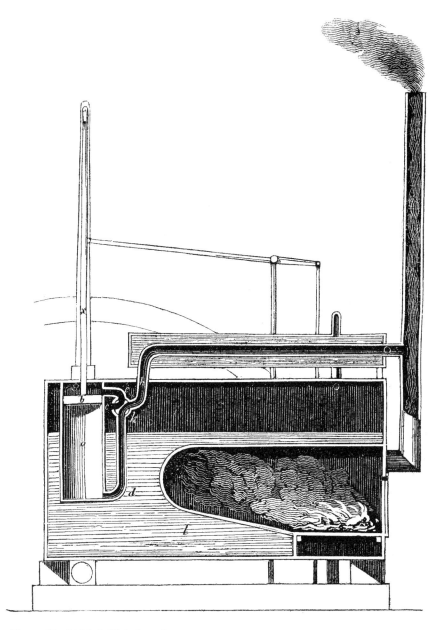

Fig. 40: Trevithick & Vivian's engine

are usually fixed in practice. *a,* is the steam cylinder; *b,* the piston; and *x,* its rod. The pipe *d, e,* connects the top and bottom; *c,* admits steam from the boiler; *b,* conveys the steam after it has moved the piston into the chimney. *k* a fourway-cock, which is moved by a lever acted on by the rise and fall of the piston-rod; *l,* the boiler in which the cylinder is placed, and the fireplace made within it, as in some other engines.

If the steam above the water is of sufficient temperature, the cock is turned into the position shown in the Figure. A communication, it will be apparent, is now open between the upper side of the steam piston and the steam in the boiler, and between the under-side of the steam piston and the chimney, or the atmosphere. The steam, however, is capable of a temperature considerably above that required merely to balance the piston, or about fourteen and a half pounds to each square inch; it would, as used in these engines, generally balance four or five times that weight; and of course the piston is impelled downwards in the cylinder. When it has made its stroke, the cock is turned by the lever or pin attached to the working-beam, or horizontal arm, into the opposite position. This opens a communication between the under side of the piston and the boiler, and between the upper side and the chimney. The preponderance of the pressures is now beneath the piston, and it is forced upwards; while the steam in the upper part of the cylinder escapes into the chimney and thus alternately. The steam, after it leaves the cylinder, is conducted through the pipe *o,* into the chimney. This pipe is inclosed in another, containing the water which is to supply the boiler by a small force pump; and from passing along this pipe it is injected into the boiler at a high temperature, which saves so much of the heat that would otherwise escape into the atmosphere and be lost.

The reciprocating motion may be converted into a rotatory one by a crank, and the motion equalized by the intermediate mechanism of a fly. The admission of steam may also be regulated by a governor and other means: but, as they have already been noticed in the description of previous engines, they are here omitted. The mechanism for ensuring the safety of the boilers is similar to that in use for condensing Engines: a safety-valve loaded with a weight equal to the pressure the apparatus can sustain; a plug of lead or other metal inserted into the side of the boiler, which will melt when the water in that vessel is heated to a certain point, or when it may have fallen below the assigned level; and the mercurial syphon of a length proportioned to the pressure,—are all so many contrivances, which, with the most ordinary care and attention,[124] are amply sufficient to give equal security to high-pressure as to low pressure boilers. We are not, however, yet in possession of any accurate experiments, by which engines constructed to raise, for instance, the same quantity of water can be compared together; but the usual performance of Trevithick's engine may be taken as equal to four-fifths of the average performance of the condensing Engines with the same quantity of coal. Placing the cylinder in the inside of the boiler saves the expense of a jacket, and effectually keeps up the temperature of the cylinder to that of the steam. All the American steam boats (except one or two) are propelled by high-pressure engines, in many cases working at double the elasticity recommended by Trevithick; yet, from the operation of a vulgar prejudice, he would be a bold speculator who should use them in an English Steam Boat, in competition with a common condensing engine, although equally safe, and more convenient from their portability and the great facility they offer of adjusting the power to the resistance, in cases where the work or load may be variable.

Steam of a high temperature was applied in a somewhat different manner, by Mr Arthur Woolfe in 1804. He had ascertained from experiment, that steam of an elasticity greater than that of the atmosphere, was capable of expanding itself as many times as its pressure was above that of the atmosphere in pound weights, and still be equal to balance the air's pressure. If, for example, one cubic foot of steam were heated so as to balance three pounds per inch on the safety-valve, it would expand itself into three cubic feet of steam, and still be equal to the pressure of the atmosphere. Steam of four pounds per inch (about 220½ degrees of temperature) would expand into a space of four cubit feet, and still be equal to balance the atmospheric column.

> At 5 pounds per square inch about 227½° it would fill 5 cubic feet.
> At 6 pounds per square inch about 230° it would fill 6 cubic feet.
> At 9 pounds per square inch about 237½° it would fill 9 cubic feet.
> At 20 pounds per square inch about 259½° it would fill 20 cubic feet.
> At 40 pounds per square inch about 282° it would fill 40 cubic feet.

Availing himself of this fact, he proposed an engine similar in its details to Mr Watt's condensing Engine, but with a second cylinder like Dr Falck's and Hornblower's. These cylinders were to be proportioned to each other, and to the elasticity of the steam which was to be introduced into them:—the first cylinder, for instance, was made to contain two or three cubic feet of steam at four pounds per inch; the second cylinder was made (if it worked with a condenser) to contain twelve cubic feet of steam, and so on. The arrangement of the parts differed nothing from that which we have described as constructed from Hornblower, except in using steam of a much higher temperature. In practice, however, it has been found that the best proportion between the cylinders is to allow of an expansion of from six to nine times—the rate of expansion as stated by Mr Woolfe; above nine times being considered (at present) somewhat problematical.

The boiler used by Mr Woolfe is similar in its principle to that which we have shown attached to Blakey's Engine: he introduces the water into a collection of pipes, placed in a furnace, and connects them with a cylindrical vessel of a larger diameter, which is attached by the pipe to the steam cylinder. Water is forced into this larger vessel by a pump. The whole arrangement of the fireplace and boiler is exceedingly favourable for procuring the greatest possible quantity of heat from the fuel: and the precautions he has taken to avoid accidents, as far as we can judge from our experience, render their use as safe as those on Mr Watt's construction. As we are not aware of any direct experiments having been made on a common condensing engine and one, with two cylinders in the *same working trim*, using the same fuel, and under the same circumstances in every other respect, it would be premature to give an opinion on the merits of the two arrangements. But it appears from published accounts of the average performance of several of the Cornwall engines, that a considerable saving of fuel arises from using the second cylinder. The variation, however, of these averages of the work of the *same* engines, would lead to the supposition of the existence of some sources of loss of effect, totally independent of the adoption of two cylinders, or of the use of one cylinder. In one average of the performance of eight condensing engines, the work is stated as equal to fifteen and three quarters millions of pounds raised a foot high with one bushel of coals; with greater care on the part of the engine-men, in another year the average was raised somewhat more than twenty millions of pounds the same height with the same quantity of coals. In the same engine the average for three months rose from twenty-two to twenty-nine and a quarter and thirty-two millions. The average performance of Woolfe's is stated at from forty-four to fifty-one millions,—in one instance nearly fifty-seven millions!

Figure 30 represents the two cylinders used by Mr Woolfe: the lever-beam, mode of opening the cocks, and other details, present nothing novel from other engines.

The small cylinder A, is connected at top and bottom with the large cylinder B; the top of the one communicating with the bottom of, the other by a pipe. The top and bottom of the large cylinder also communicate by the pipe Y, with the condenser: the pipe *c,* conducts steam from the boiler into the small cylinder—both are closed at top and bottom. The pistons work through stuffing boxes, and the precautions taken in the common condensing engine to keep the cylinders as hot as the steam that enters them, are observed in this engine; and the manner of condensing the steam and freeing the cylinder from the injection water and air, is the same as in the common engines. We will suppose the pistons to have made a stroke, and that they are at the top of their cylinders: in this case the communication between the bottom of the small cylinder and the top of the large one is open; the bottom of the large cylinder will also be open to the condenser, and the steam has liberty to flow from the boiler on the top or the smaller piston: the elastic force of the steam above this piston pressing it downwards, while the steam beneath, being allowed to flow into the larger cylinder by its expansive force, will impel its piston into the vacuum made under it. By shutting all these cocks, and opening those which connect the bottom of the small cylinder and the boiler, and the top with the bottom of the large cylinder, while the passage from the condenser is open to the upper side of the larger piston the motion will be reversed.[125]

Mr Hornblower's second steam-wheel acts on the same principle with one he had constructed a few years before, but is much simpler in its arrangement. It has a rotatory motion within itself by the action of four revolving pistons. The pistons are four vanes like those of a smoke jack, but of iron of some considerable thickness, sufficient to form a groove to hold some stuffing for the purpose of being tight in their action. They are mounted on an arbour, which has a hollow nave in the middle; the tails of the vanes are inserted into this nave, and each opposite vane is affected alike by having a steady connexion with each other; so, should the, angle of one of the vanes be altered, the opposite one will be altered too: and they are not set in the, same plane, but at right angles to each other. If we conceive these vanes to be held in a vertical position, like the sails of a windmill, when one vane is flatly opposed to the wind, the opposite vane will present its edge to it; and this they are constantly doing in their rotation on their common arbour, and the steam acts against the vane on its face in propelling it into action for about the quarter of a circle, or 90°, in the box wherein it acts, and then it instantly turns its edge to the steam, while at the same time another vane has entered the working part of the box, and the rotation proceeds without interruption.

Figure 41 is a horizontal section of the engine. *a*, the outside box or casing; *b, c*, two of the vanes, *c*, in the position of presenting its whole surface to the steam, and *b,* its edge; *d,* the naves in which the vanes are connected; *o*, the arbour. Figure B, is a vertical section of this steam-wheel, and the same letters in both Figures refer to the same parts. The nave in the middle shows how

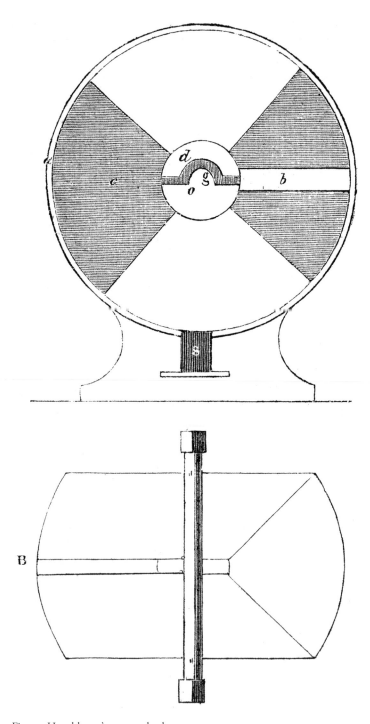

Fig. 41: Hornblower's steam wheel

the vanes are connected; the other pair being joined in the same manner, will have liberty to turn a quarter round without obstruction by the crooked part of the communicating axis. *s,* is the steam pipe. The exhausting pipe or condenser, and discharging pump, may be attached to the apparatus in the usual manner.[126]

At the beginning of this century not more than "four engines of any importance" were at work in the whole continent of America: of these one supplied New York with water, another gave motion to a saw-mill, and the other two belonged to the corporation of Philadelphia[127]

In 1804 Mr William Lushington introduced one of Mr Watt's engines into the colony of Trinidad, and we believe that the expense of its work was estimated at about one third only of the same quantity of labour performed by the cattle-mills commonly used in the colony.[128]

An ingenious combination of Savery's and Papin's apparatus was proposed in 1805, by Mr James Boaz of Glasgow. Figure 42 shows one of his arrangements, by which water can be raised without condensing the steam. *a,* is the steam cylinder; *i,* the pipe from the boiler, having a stop-cock; *k,* a waste steam cock; *e,* a floating piston attached to a piston-rod. E a pipe which generally contains hot water; *f* water pipe, having a valve at *g* immersed in the well, and delivering the water into the reservoir *o,* through a valve, *z.* The air which accumulates in the receiver escapes at *n; o,* the raised water cistern; *d,* rarifying or exhausting vessel.

The whole being filled with mercury and water, shut the air-valve *s,* and open *i;* the steam from the boiler will rush into the receiver, and, after heating the water, it presses on its surface, forcing the mercury up into the exhausting vessel *d,* where it is shown in the engraving by a darker shade. The water above *c,* and in the pipes. *e, f,* will be forced up into the cistern *v,* in a quantity nearly equal to the space occupied by the steam in the receiver. When the piston has been depressed as far as is necessary for its stroke, the self-acting mechanism attached to its rod, shuts *i,* and opens *k*; and the mercury now being at liberty to act by its gravity, descends from the exhausting pipe, and raises the piston to its first position; and the steam which pressed it downwards being now allowed to flow into the atmosphere, the fall of the mercury from *d,* into *a,* leaves a vacuum in *d,* into which the water from the well is pressed by the atmosphere, and again fills it. The valve at *g,* prevents its return to the well in the operation of forcing; and the valve *z,* prevents its fall from the cistern when the vacuum is made in *d*[129]

The rotative Engine (Figure 43,) constructed by Mr Andrew Flint, consists of two concentric cylinders or drums, placed at a certain distance from each

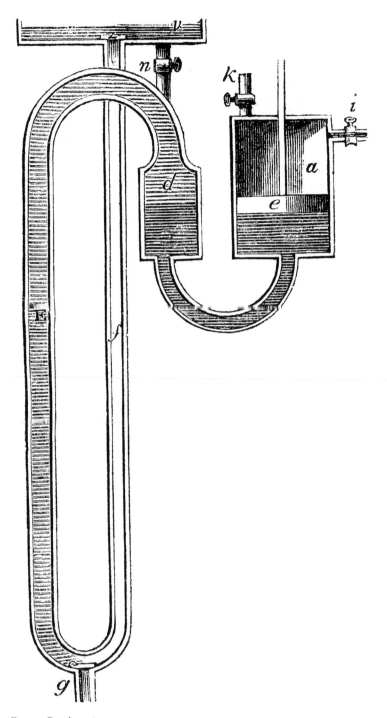

Fig. 42: Boaz's engine

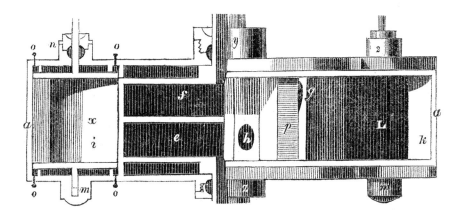

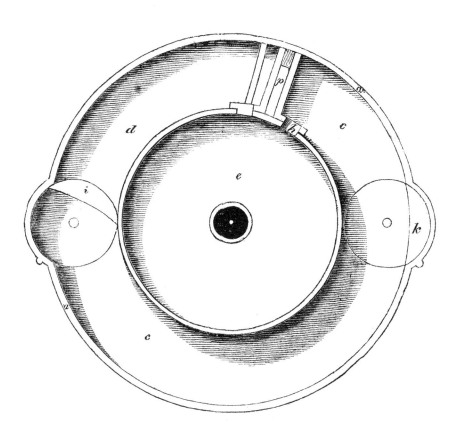

Fig. 43: Flint's rotatory engine

other. To the inner cylinder is fixed a piston *p*, moving and made steam-tight in the space x, between the cylinders; and it is attached to two hollow axes *y, z*. The inner cylinder is divided in its height into two chambers, *e, f;* having two openings *g, h. i, k*, are two valves, which are moved alternately, by levers acted on on the outside from the revolution of a spanner or lever attached, to the hollow axle, one being always open when the other is shut.

The valve *i,* in the plan, being shut, and *k* open, and the piston in the situation shown in the Figure, the steam issues through *h*, and impels the piston towards *k;* the air which may be in the space on the other side of the piston, escaping through *g*, into the chamber *f,* which communicates with the condenser. When the piston has passed valve *k,* it is shut; the steam will then be confined to the space between it and the piston, and will continue to impel the piston forward; the air still escaping as before on the other side of the piston through the movable aperture *g*, so that the quantity of steam impelling the piston can never be more than half the contents of the channel of revolution; it is then admitted into the vacuum made on the other side of the piston, and drawn off by the condenser. The idea is ingenious; but the friction must be enormous. Mr Robert Willcox's Engine,[130] differs from Mr Flint's only in the mode of placing the cocks, and the position of the pipes from the boiler and condenser.

In his rotative Engine, Mr Mead, in place of the cylinders substitutes shells, or a circular ring in which the pistons move; and instead of making the apertures for the admission of steam, and its exit to the condenser, revolve with the two pistons, he makes them in a part of the fixed case, and the passage of the piston over them in each revolution is accomplished by the momentum of a fly.

Mr Richard Witty has published a description of two engines moved by steam, the second being an improvement on his first. His scheme is peculiar: "he combines the reciprocating rectilinear motion with the rotative in such a manner, that the steam cylinders with pistons moving in them in a rectilineal direction, do at the same time turn round on a horizontal axle, and partly form a fly-wheel."[131] The second patent describes that the further improvement consists in "making the piston draw or force the machinery to be worked by it, whilst itself moves both in a rectilineal and rotatory direction in a cylinder or steam vessel, which also revolves on its axis placed in any position."[132]

The first Steam-boat in America was launched at New York in October 1807, and began to ply between that city and Albany.[133]

Mr Samuel Clegg's rotative piston makes a complete revolution in a channel at a distance from the centre of motion. The details are very different from those of any that have been described, and much more likely to succeed in practice. On this principle Mr Clegg constructed several engines of a good size; and he informs us that he found them to answer all the ends he intended, by their introduction; they took up very little room, made no noise, and could be manufactured at about half the price of condensing engines.

In Figure 44 the bottom plate of this machine (supposed to be turned up) is made perfectly smooth and flat. A series of blocks are placed in a groove concentric to the outer rim of the plate; these blocks are of considerable weight, and occupy the entire circular groove excepting the space v, which is fitted with adjusting springs and screws to keep the series of blocks close together; the sides which apply to each other are ground plain, so as to make them fit, and be as nearly steam-tight as possible; the under surfaces are also made flat, and the whole series forms one flat horizontal surface, except the space x; which has the screws, and these are sunk so far as to allow a flat bar to pass clear over them. i, is the axis which communicates motion to the machinery. f, is a bar, to which the revolving steam piston is attached, having a small wheel g, fixed upon it. The movable blocks are enclosed on their two sides and top by an iron box e, d, in which they slide upward and downward. The piston moves round in a semicircular channel or chamber, having a valve or flap at n, and the steam is admitted at v. The segments are described as being perfectly flat on their under-side, but in fact a part of their under-surface is formed with a small curve, which appears from the shading on the Figure A.

When the piston is placed as in the Figure, and steam introduced between it and the valve or door at n, it will recede from the pressure, and be carried round; and the air which is in the chamber, on the opposite side of the piston, will be pressed out through the apertures x, x, into the atmosphere, or into a condenser. During the progress of the piston, the segments being in fair bearing on the flat rim of the piston chamber, will keep the space into which the steam is admitted steam-tight. The roller attached to the bar or piston rod, (being made to precede it) presses upon the under-side of each segment in its revolution and raises it just sufficient to let the thin bar (about ⅝ths of an inch thick for the largest engine), or piston rod, slide between its under surface and that of the plate or rim. The weight of each segment, after the bar has passed under it, makes it again fall into its place; or another roller may be made to follow the piston and depress the block into its situation. The passage of the bar under each block is completed before any communication can possibly be made between the external air and the front of the piston. When the steam piston arrives at the valve or door, which hangs from the flat plate covering the chamber, and moves only in one direction, the piston presses it up into a recess, and passes it. The valve,

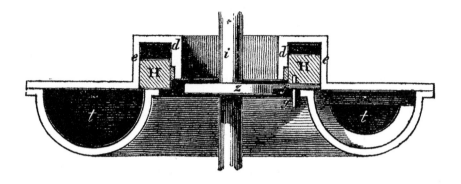
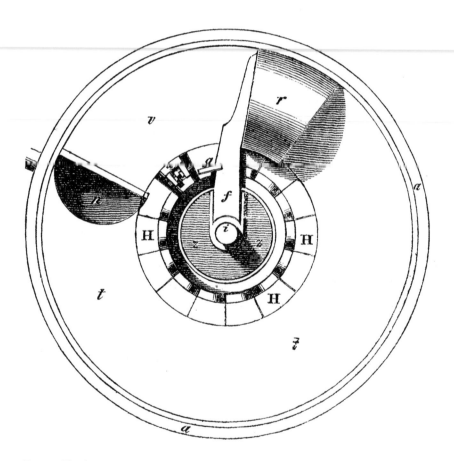

Fig. 44: Clegg's rotatory engine

admitting steam to the piston, being also at this moment shut, the piston is carried beyond the position of the hanging valve by the momentum of a fly wheel. When it has moved so much forward as to allow the valve to fall, the steam is again admitted between the valve and the piston, which makes another revolution. The steam which has propelled the piston being now in communication with the atmosphere by the apertures z, escapes into it. The flat face to which the blocks are ground, is about three inches broad, with a groove near the centre, to introduce any kind of elastic packing; but Mr Clegg found this packing was not required. In his experimental machine the shafts give motion to the air pump steam-valve in the usual manner.

Mr Clegg estimated that a piston of an engine equal to the power of twenty horses, will move through a circumference of twenty feet; each block for an engine of this size will weigh about twenty pounds, and will not be required to be raised more than half an inch or five-eighths of an inch, to allow the bar of the largest engine to pass. From the pressure upon a piston being equal to 4000 pounds weight moved twenty feet, is to be deducted the raising of these blocks, weighing 500 pounds, five-eighths of an inch; and this, Mr Clegg calculates, will be the power of his engine: but, were we to treble the estimated resistance, it would still be a very great performance.[134]

The external appearance of the apparatus proposed by Mr Turner in 1816, as a rotative engine, very much resembles that by Mr Mead; and the general arrangement of the parts and manner of action is also similar to that gentleman's; but on the whole the mechanical tact of Mr Turner is greatly superior to that of his rival, and we shall be more minute in our description of his apparatus.

The steam-piston, w, (in Figure 45,) moves in a circular channel or ring, and is firmly attached to the upright axle, f, which communicates the motion to the machinery by means of wheels and pinions. The construction of the two valves which divide the piston-chamber is shown by the Figures, and will be sufficiently understood from inspection. The machinery which moves the sliders or valves, a, b, and that which opens and shuts the valves for the admission and exit of the steam, are connected, and "act in concert:" so that, as soon as the piston w, has passed the slider or valve, it shall be lowered into its seat, and the instant after the piston has passed by the opening for admitting steam, its valve shall be opened to admit of steam being introduced between it and the piston. The edge of the centre-plate g, to which the pistons are attached, must be made to move tight between the upper and lower valves which compose the cylinder, so as to prevent the escape of steam between them; and at the same time that it may revolve

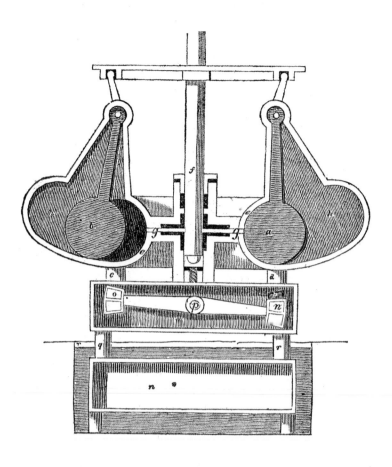
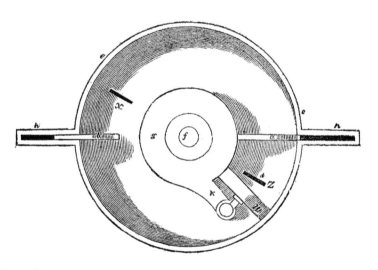

Fig. 45: Turner's rotatory engine

freely, its outer edge has a kind of projecting ring. *n, o,* in the upper figure, are the two steam-valves in the bottom of the circular passage; which have been described as being opened and shut by a mechanism which opens and shuts the sliders. *m* is an iron box, into which steam-is admitted from the boiler; two pipes*, c, d,* proceed from it into the cylinder, and other two, *q, r,* from it to the condenser. The bar or lever moving on the pivot *p,* has a box or cup at each end, capable of covering the openings of two of these pipes at once: one from the boiler and the other to the condenser being placed at each end, and the faces of the boxes are ground to fit close to the face of the steam-chest. When the lever moves up or down, it opens one of the pipes into the piston-chamber, and shuts that situated under it from communicating with the condenser. The box at the other end of the lever covers the passage for the steam, and opens a communication between the other side of the piston-chamber and the condenser. When the piston is in the situation shown in the Figure, the slider *a* is that (and the opposite slider is drawn up), or in the position shown in the upper figure; and steam is admitted between it and the piston by the opening communicating with that marked with the letter *z*, in the under figure. It is then ampelled towards *b,* and the air or steam being drawn off by the condenser through *r,* a vacuum is formed on the other side. When the piston has reached *b,* the slider is drawn up, and as it passes onwards *b* is lowered; and the piston being carried beyond and over the opening *x,* the box in the steam-chest shuts off its communication with the pipe leading to the condenser, and opens it to the steam-chest: at the same instant reversing the openings on the other side; by which steam is shut off from *o,* and its communication is effected with the condenser. A vacuum is now made on the side of the piston opposite *w,* when the elasticity of the steam introduced from *x,* between *b* and *x,* carries it forward. After it has passed the valve *a,* it is again lowered, and *b* is opened; at the same time steam issues between *a* and *w,* and *o* is again connected with the condenser: which produces a vacuum on one side, and a pressure on the other, giving a rotatory motion.[135]

Amonton's idea of producing a rotatory motion by the pressure of a column of water placed on one side of a wheel, was made the basis of a most ingenious mechanism by Mr William Onions in 1811. The steam was introduced into the axis of the wheel, and allowed to flow by the radiating arms into chambers formed in the circumference. A vacuum being alternately formed in these chambers, water was allowed to rise by the atmospheric pressure, which by its gravity produced a rotatory motion. Many of the details in this engine are excellent; but as a whole, it was considered to be too complex for its ingenious inventor to entertain very sanguine hopes of introducing it into practice.

The first attempt on any scale worthy of notice at navigation by steam in Britain, was made about this period on the River Clyde. A boat of about forty feet keel and ten and a half feet beam, having a Steam Engine of three-horse power, began to ply on the Clyde as a passage-boat between the city of Glasgow and Greenock in 1812[136]; but owing to the novelty and apparent danger of the conveyance, the number of passengers was so very small, that the projectors for some time hardly cleared their expenses!

The introduction of the Steam Engine into Peru arose from a series of incidents which have almost the air of romance. We have described the High Pressure Engine of Mr Richard Trevithick, for which he obtained a patent in 1802. In order to give a better idea of the arrangement and operation of his engine, he had constructed a working model, which was so highly finished that it had found its way to London as a cabinet curiosity.

About the same period M. François Uvillé had found in Peru some of the richest mines falling into decay, or totally drowned, from the impossibility of draining them by manual labour; and learning also that these mines were considered to be richer in silver ore than those of Mexico, he conceived the idea of introducing the Steam Engine as a substitute for animal power, to accomplish their draining. M. Uvillé came to London in 1811; but, among those whose opinions he asked, he met with no encouragement to pursue his plan, on account of the inefficacy of steam in an atmosphere so rare as that in which the Peruvian mines are situated among the Cordilleras, as well as the seeming impracticability of conveying the parts of those large engines which would be required, over mountains inaccessible to any species of wheel-carriage.

About to leave England in despair of ever being able to accomplish his grand project, and passing by a street leading from Fitzroy-square, he accidentally saw a model of a Steam Engine exposed for sale in the shop of a Mr Roland. He examined it with great attention, and, being struck with the simplicity and excellence of its principle and construction, he became its purchaser at a price of twenty guineas—this was the Trevithick model.

M. Uvillé felt he now had in his hands the means, either of carrying forward his gigantic scheme—or of setting the fever of his mind at rest, by demonstrating the impossibility of achieving his project by the medium of the Steam Engine. He carried the model to Lima, and lost no time in trying its effects on the highest ridges of Pasco, which form the site of the mines: to his unspeakable joy the experiment succeeded to his most sanguine wish, and to that of some others who witnessed it. And in July 1812, he formed an association with Don Pedro Abadia and Don Jose Arismendi, opulent merchants of Lima, for the purpose of contracting with the proprietors of the flooded mines. The Marquis de Concordia, then Viceroy of Lima, highly approving the plan; under his protection the new Company succeeded in getting contracts to work several of the principal mines for the payment of about a fourth part of the produce which they might bring to the surface. These contracts were made in August 1812; and in pursuance of his scheme, M. Uvillé again embarked for Europe, and reaching Jamaica, he took his passage for Falmouth. M. Uvillé's mind was too full of the flattering expectations which his scheme inspired, not to be making frequent inquiries among his fellow passengers about mines and engines. One day conversing with a Mr Teague, and expressing an anxious wish to find out, if possible, the author of the model he had carried to Lima, he was most agreeably surprised to hear Mr Teague reply, "Mr Trevithick is my near relative, and within a few hours after our arrival at Falmouth I can bring you together." It happened accordingly; and M. Uvillé resided several months at Camborne with Captain Trevithick, receiving during his stay instructions from that able man in mining and the management and construction of machinery. Accompanied by Captain Trevithick, he visited other mining districts; and being introduced to Messrs Bolton and Watt; the first steam-engineers in the universe, he explained to these gentlemen the mountainous precipices to be surmounted, and the great elevation of the mines above the level of the sea. But whether the objection of these gentlemen to engage in a speculation in which there was much uncertainty of carrying it forward with effect, either from the disturbed political state of Peru, or the difficulty of transporting the parts, or the great elevation of the situation being adverse to a favourable employment of condensing engines, does not appear; but their opinion was unfavourable to M. Uvillé's project.

Captain Trevithick and his friend were not, however, to be discouraged from making the attempt; and in January 1814 the Captain entered into an engagement with M. Uvillé to provide nine engines, at a cost of about ten thousand pounds; which by very great exertions were shipped, with the permission of the British government, from Portsmouth in the September following, accompanied by M. Uvillé and three Cornish men to direct the erection of the machinery. On arriving safely at Lima, they were welcomed by a royal salute and public rejoicings. But after they had got so far so great were the local obstacles in transporting the heavy masses across the mountains, that it was not till the middle of 1816 that they were able to set one of the engines to work. This was the first ever seen in South America, and excited intense curiosity. Great ceremony, it appears, was observed on this important occasion; and the most distinguished honours were conferred on the projectors by the Vice-Regal government. The official deputation appointed to superintend this new and very extraordinary operation, made a report to the Viceroy, which was published in the *Lima Gazette* in August 1816. "Immense and incessant labours," say the reporters, "and boundless expense have conquered difficulties hitherto deemed insuperable; and we have, with unlimited admiration, witnessed the erection and astonishing operation of the first Steam Engine. It is established in the royal mineral territory of Taüricocha in the province of Tarma; and we have had the felicity of seeing the drain of the first shaft in the Santa Rosa mine, in the noble district of Pasco. We are ambitious of transmitting to posterity," they continue, "the details of an undertaking of such prodigious magnitude, from which we anticipate a torrent of silver that shall fill surrounding nations with astonishment."

They then go on to name a number of individuals on whom "the eternal gratitude of all Spaniards is invoked;" and it is somewhat remarkable that the only Englishman mentioned by name in this list of worthies is *Mr Bull*.[137]

While these operations were going on in Peru, Captain Trevithick in England was vigorously engaged in providing further supplies of Steam Engines, constructing coining apparatus for the Peruvian mint, and in constructing furnaces for purifying the silver ore by fusion: a project of incalculable importance from the increasing scarcity of quicksilver. This second supply was sent from England in October 1816, and arrived at Lima in the February following. Captain Trevithick went out in this vessel. On his arrival he was immediately presented to the Viceroy, and most graciously received, and his arrival officially announced in the Lima Gazette. Public notice was at the same time given in this of the completion of the second engine, said to be superior in power and beauty to the first; and also of the reception of some parcels of ore of extraordinary richness, raised from the mines restored to use by the operation of these machines. The Gazette also announces the arrival of the other engines; "but that," it continues, "which is of still greater importance is the arrival of DON RICARDO TREVITHICK, an eminent professor of mechanics, machinery, and mineralogy, inventor and constructor of the engines of the last patent, and who directed in England the execution of the machinery now at work in Pasco. This professor, with the assistance of workmen who accompany him, can construct as many engines as shall be wanted in Peru without the necessity of sending to Europe for any part of these vast machines. The excellent character of DON RICARDO, and his ardent desires to promote the interests of Peru, recommend him to the highest degree of public estimation, and make us hope that his arrival in this kingdom will form the epoch of its prosperity through the enjoyment of its internal riches, which could not be realized without such assistance; or if the British government had not permitted the exportation from England: an object hitherto deemed unattainable by all who know how jealous that nation is of all her superior inventions in the arts or Industry."

So much importance was attached to Don Ricardo Trevithick's personal superintendence, that the Viceroy ordered the Lord Warden of the Mines to escort him with a guard of honour to the mining district, where the news of his arrival in Peru caused the greatest rejoicings; and many of the chief men came to Lima, a distance of many days' journey over the mountains, to welcome him. M. Uvillé had written to his associates "that Heaven had sent him out for the prosperity of the mines, and that the Lord Warden had proposed to *erect his statue in massy silver*."

In this narrative of romantic incidents it should not be forgotten that Don Ricardo is now superintending the royal mint of Peru; and so strong are the hopes of the success of his operations, that he has had orders to augment the powers of the *coining* machinery *six-fold*. The latest accounts, according to Mr Boase, left Don Ricardo in the enjoyment of increasing distinction, and a flattering prospect of great wealth. In addition to his emoluments as patentee and engineer, he has the fifth part or share in the Lima Company, which, at a moderate estimate, amounts to one hundred thousand pounds a-year.[138]

Sir Wm. Congreve's wheel is similar to that which we have described as suggested by Mr Watt. His apparatus may be explained as resembling a water-wheel moving in water. The steam entering into the boxes at the lower portion of its circumference, by its superior elasticity it raises them to the surface, and produces a continuous motion round an axis. No description, however, of this scheme is published.

Since his retirement from active life, in 1800, Mr Watt[139] continued to reside at Heathfield near Birmingham, and had the rare felicity of feeling that the detraction and envy of some of his contemporaries had long disappeared in the homage paid by his countrymen to his virtue and genius, and of seeing the introduction of his inventions into every manufacture, and to note the prodigious extension which by this means was given to English commerce; he lived to witness the triumph of his country, and to hear his inventions estimated as being mainly instrumental in the formation of that national strength and influence which led to the glorious consummation. Reverenced by his family and friends for the benevolence and warmth of his affection, full of years, this venerable man closed a life illustrious from its usefulness, on the 23d of August, 1819, more from a decay of nature than any particular disorder.

His name fortunately needs no commemoration of ours, for he that bore it survived to see it crowned with undisputed and unenvied honours; and many generations will probably pass away before it shall have gathered all its fame. We have said that Mr Watt was the great improver of the Steam Engine; but in truth, as to all that is admirable in its structure, or vast in its utility, he should rather be described as its inventor. It was by his invention that its action was so regulated as to make it capable of being applied to the finest and most delicate manufactures, and its power so encreased as to set weight and solidity at defiance. By his admirable contrivances, it has become a thing alike stupendous for its force and its flexibility; for the prodigious powers which it can exert, and the ease and precision and ductility with which they can be varied, distributed, and applied. The trunk of an elephant that can pick up a pin or rend an oak, is nothing to it: it can engrave a seal, and crush masses of obdurate metal like wax before it; draws out without breaking a thread as fine as a gossamer, and lift a ship of war like a bauble in the air. It can embroider muslin, forge anchors, cut steel into ribands, and impel loaded vessels against the fury of the winds and waves.

It would be difficult to estimate the value of the benefits which these inventions have conferred upon the country. There is no branch of industry that has not been indebted to them, and in all the most material, they have not only widened most magnificently the field of its exertions; but multiplied a thousand fold the amount of its productions. It is our improved Steam Engine that has fought the battles of Europe, and exalted and sustained through the late tremendous contest, the political greatness of our land. It is the same great power which now enables us to pay the interest of our debt, and to maintain the arduous struggle in which we are still engaged, against the skill and capital of all other countries. But these are poor and narrow views of its importance. It has increased indefinitely the mass of human comforts and enjoyments, and rendered cheap and accessible, all over the world, the materials of wealth and prosperity. It has armed the feeble hand of man, in short, with a power to which no limits can be assigned, completed the dominion of mind over matter, and laid a sure foundation for all those future miracles of mechanic power, which are to aid and reward the labour of after generations. It is to the genius of one man too, that all this is mainly owing, and certainly no man ever before bestowed such a gift on his kind: the blessing is not only universal, but unbounded; and the fabled inventors of the plough and the loom, who were deified by the erring gratitude of their rude contemporaries, conferred less important benefits on mankind, than the inventor of our present Steam Engine.

This will be the fame of Watt with future generations, and it is sufficient for his race and country; but to those to whom he more immediately belonged, who lived in his society and enjoyed his conversation, it is not perhaps the character in which he will be most frequently recalled, most deeply lamented, or even most highly admired.

No man could be more social in his spirit; less assuming or fastidious in his manners, or more kind and indulgent towards all who approached him; his talk, though overflowing with information, was full of colloquial spirit and pleasure. He had a certain quiet and grave humour, which ran through most of his conversation, and a vein of temperate jocularity, which gave infinite zest and effect to the condensed and inexhaustible information which formed its main staple and characteristic. His voice was deep and powerful, though he commonly spoke in a low and somewhat monotonous tone, which harmonized admirably with the weight and brevity of his observations, and set off to the greatest advantage the pleasant anecdotes, which he delivered with the same grave brow and the same calm smile playing soberly on his lips. He had in his character the utmost abhorrence for all sorts of forwardness, parade, and pretensions, and indeed, never failed to put such impostors out of countenance, by the manly plainness and honest intrepidity of his language and deportment. In his temper and disposition he was not only kind and affectionate, but generous and considerate of the feelings of all around him, and gave the most liberal assistance and encouragement to all young persons who showed any indications of talent, or applied to him for patronage or advice. His health, which was delicate from his youth upwards, seemed to become firmer as he advanced in years; and he possessed up to the last moment of his existence, not only the full command of his extraordinary intellect, but all the alacrity of spirit, and the social gaiety which had illuminated his happiest days.

His happy and useful life came at last to a gentle close. He had suffered some inconvenience through the summer, but was not seriously indisposed till within a few weeks from his death. He then became perfectly aware of the event which was approaching; and with his usual tranquillity and benevolence of nature, seemed only anxious to point out to the friends around him, the many sources of consolation which were afforded, by the circumstances under which it was about to take place. He expressed his sincere gratitude to Providence for the length of days with which he had been blessed, and his exemption from most of the infirmities of age, as well as for the calm and cheerful evening of life, that he had been permitted to enjoy after the honourable labours of the day had been concluded. And thus, full of years and honour, in all calmness and tranquillity, he yielded up his soul without a pang or a struggle, and passed from the bosom of his family to that of his God.[140]

Mr Job Rider's rotatory Engine for which he obtained a patent in 1820, operates by the elasticity of steam (acting against a vacuum,) on a series of valves placed on an axis which communicates the rotatory motion, in a manner similar to Cooke's and Cartwright's. But Mr Rider's valves (and these are the objectionable parts of all the others) are constructed in a very ingenious manner, and the apparatus may be considered to be more practical than any of those we have noticed as being constructed on this principle. In Ireland Mr Rider has erected some engines of large dimension, and has worked them for that length of time which enables him to speak with certainty of the effect of his machine. As it might be considered a needless addition to the size and expense of this book to figure and describe an apparatus which has been given in the 'Mechanic's Magazine,' a book in the hands of every mechanic, we refer to the first volume of that meritorious periodical for ample details, and a good engraving of Mr Rider's invention.

In Mr Moore's scheme, the rotatory motion is produced by the revolution of a wheel, with floats or vanes on its circumference, moving on hinges. a, is the wheel, having projections or boxes in its circumference for the reception of the valves or "propellers" d, d, d, d. This wheel and valves move in a fixed cylinder, inclosed at the ends so as to form the space between the moveable drum, and the fixed case, air and steam tight. Angular pieces x, x, are fixed to the outer case and sides, and make a fixed steam-tight division of the circumference. Steam is admitted by the pipes h, e, e, from the boiler, and the pipes g, f communicate with the condenser. The valves d, move on an axis, and are opened and shut by levers attached to it on the outside. In the position of the wheel shown in the engraving, steam is entering from the boiler through h at e, e; the valves are shut; and the communication being open between the spaces shown with a dark shade and the condenser, the steam presses the pistons or valves forward in that direction. When they arrive at the angular pieces, which block up the communication between the semi-circumferences of the wheel, they are moved upon their axes into the boxes, as shown in Figure 46, and are carried forward until they pass the fixed part attached to the angular pieces x, x, in order to make a steam-tight passage, and, emerging at the other side, they are again turned to fill up the steam chamber. When pressed on one side by steam from the boiler, with a vacuum on the other, a very powerful continuous motion is generated. Mr Moore's idea is a very

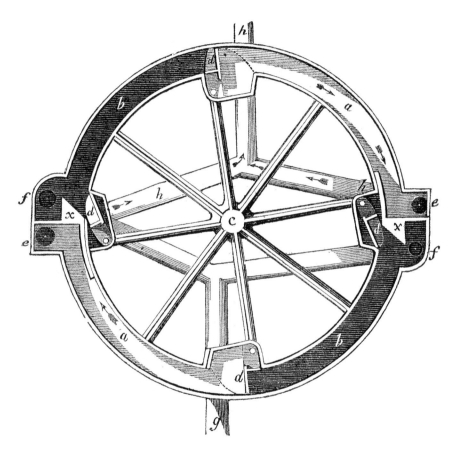

Fig. 46: Moore's rotatory engine

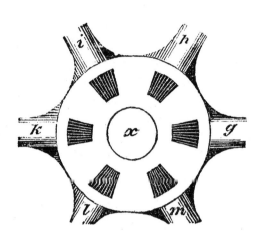

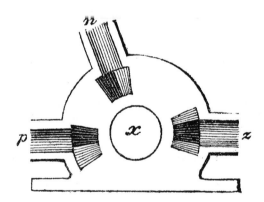

ingenious one; and, could a mode be suggested of making the inner revolving drum steam-tight, without so great a loss of power as takes place in its present form, from the friction of so large a surface as the edges of his revolving cylinder, it appears capable of considerable precision in its action.[141]

The most simple and practical of all the projects for producing a rotatory motion by the weight of a column of fluid acting on the circumference of a wheel, is that proposed by Mr Thomas Masterman in 1820.[142] In principle and arrangement, it is similar to Mr Onions', but in everything relative to simplicity of construction, convenience of arrangement, and effective action, it is beyond comparison better adapted to practice.

Figures 47 and 48 are vertical sections of this Steam Wheel. *a, b, c, d, e, f,* are weights at the ends of short levers, which open and shut the valves during the revolution of the wheel. *s, s,* is a steam-tight ring, divided into six or more compartments by the valves attached to the levers. Each of these chambers has a communication with a series of perforations surrounding the axis of the wheel, by the radiating arms or channels *g, h, i, k, l, m.* This series of perforations revolves with the wheel against a fixed plate, which has three openings or perforations: one of these openings, *p,* leads to the condenser; *n,* to the boiler; and *x,* to the water cistern. *x,* is the axis on which the wheel turns, and by which motion is communicated to the other machinery. The valves *a, b, c, &c.* are moveable on a joint or hinge, and are connected by a spindle working steam-tight through the sides of the rim of the wheel, and attached to the weights already noticed.

The moveable series of perforations revolving with the wheel, are brought in rotation opposite each of the three perforations in the fixed plate; and each of the arms, and chambers in the circumference of the wheel, communicates in turn with the condenser, water column, and boiler. In the position of the wheel shown in the Figures, the moveable opening of arm is opposite to the perforation *p,* which leads to the condenser; arm *l,* will then communicate with the water column by the perforation *n.* When steam from the boiler is allowed to flow through *p,* it will fill that part of the circumference between *h,* and *g,* which by its elasticity will prevent the water front rising on the side in which it is admitted; but it will (if made of greater elasticity) have a tendency to force the water upwards on the other side, from the valves being made to open in that direction only. The pipe *h,* communicating with the condenser, the column of water on the circumference of the wheel, being relieved from the pressure of the atmosphere, will rise to a height corresponding to the temperature of the condenser; this gives a preponderance to that

Fig. 47: Masterman's rotatory engine

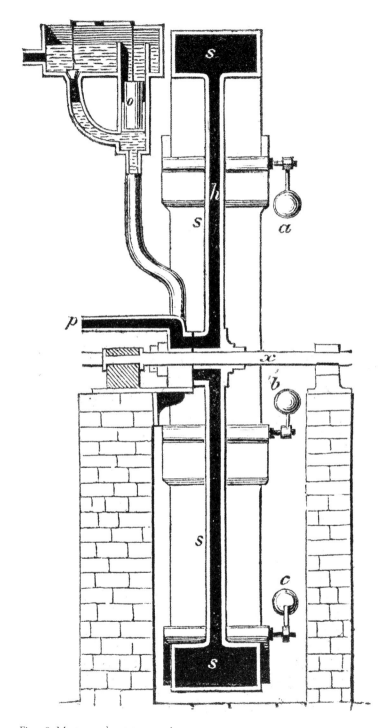

Fig. 48: Masterman's rotatory engine

side of the wheel, which brings other arms into the same position, so that a continuous motion round the axis x, is generated.

In this modification of the principle the moving power is derived from the gravity of the water; but although the column of fluid (when the condenser is used) cannot exceed in height a column equal to the atmospheric pressure, its horizontal section may be made of any dimensions. In practice, however, it is recommended by the inventor not to work it with a less pressure than about twenty-eight feet. It is obvious that this wheel may be worked by high-pressure steam, by merely allowing the pipe which leads from the fixed perforation to open into the atmosphere instead of opening into the condenser, and the wheel may then be made of any diameter; and it is estimated that it might be worked in this manner with about half the pressure of steam of a reciprocating engine.

Mr Masterman states (from experiment) that his wheel is a great deal cheaper than a common condensing lever engine, costs less to work and repair, occupies a smaller space,—possesses a greater facility and certainty of being put in motion, besides realizing a great saving of power from the small friction of its parts compared with that on other constructions. On the last point an experiment is stated in which the friction of a wheel fifteen feet in diameter, and weighing about three tons, was kept in motion by a pressure of three-eighths of a pound per square inch of horizontal section of the water in the ring, "ascertained by a glass mercury gauge attached to the steam-pipe;" and as the valves were about "78.6 square inches, this was not more than 30 pounds on the closed valve."[143] This wheel revolved with a speed of 688 feet per minute, with a pressure of about three pounds and a half per inch; but a considerable part of this pressure must have been occasioned by the confinement or wiredrawing of the steam in passing through the radii: the most economical speed, however, is about 400 feet per minute.

Were the diameter of the wheel to be about twenty-eight feet, the depth of the rim twelve inches, with the radiating pipe or arms three and one-eighth inches, the whole turning on an axis of six inches in diameter, with a column of water equal to ten pounds on the inch, and using a condenser, this will be a twelve-horse-power engine; and with a column of water equal to ten pounds acting against the atmosphere, it will be equal to a power of thirteen horses; but the expenditure of fuel will be greater in proportion to the effect, in the ratio of about 150 to 94. In comparing the effect of a reciprocating engine working with a pressure of seventeen pounds per square inch, and leaving only an available power of seven pounds, Mr Masterman obtains an available power of eight and one-sixth pounds weight from a pressure of thirteen and a half pounds, so that the effect from the same quantity of coal will be about 150 for the rotatory engine and 95 for the common condensing engine. We have been thus minute in our account of this apparatus, from the wish to excite attention among mechanics to the mechanism of rotatory engines; should these ever be brought into successful competition with reciprocating engines, it appears probable that they may be constructed on the principle adopted by Mr Onions and by Mr Masterman. Mr Masterman, however, as an individual (keeping the merit of his Steam Wheel out of the question) is entitled to that approbation which is due to liberality of feeling and conduct. The ample description of his machine with an illustrative engraving, and the fair and candid appeal he makes to experiment, in the tract he published in 1822, is worthy of all praise;—were this practice more frequent among mechanics, especially inventors and patentees, the advantages to themselves, and ultimately to the public, would be incalculable. Mystery in anything is bad; and that invention which is not placed in all its details before the public, has not seldom been so withheld from a latent feeling in the patentee either of its want of originality or of its want of merit.[144]

From the first invention of the Steam Engine, and in all the combinations and forms of the boiler, the steam, of whatever elasticity, was always generated in contact with the water. This practice demanded vessels of considerable dimension, as the boilers were obliged to be proportioned in their capacity to act also as reservoirs of steam. Which necessarily limited the engineers to the use of vapour of comparatively low temperature; for it was next to an impossibility to construct a very large vessel of sufficient strength to be safe from the danger of explosion, if containing steam of a great elasticity. It had long been known, also, that water might be raised to a very high temperature by preventing the formation of vapour by pressure—that it could be even made red hot; the difficulty lay in doing this.

As water can be subjected to almost any assignable pressure by a forcing pump, it occurred to Mr Jacob Perkins, that it might also be heated with facility under any pressure so produced; it might be heated in one vessel, and then be forced from it into another, in which it might expand into vapour, and be thus made available to every purpose for which steam is applied at present. His contrivance for doing this was exceedingly ingenious; and, in 1828, after he had obtained a patent, he exhibited to the public a large working model of his invention.

This experimental apparatus had a boiler made of copper about three inches thick, and of a capacity to hold eight gallons of water. It was closed at the lower end, and at the upper end it had five small perforations into which were inserted as many pipes. This cylinder was placed vertically in a furnace in which the fuel was kept in vivid combustion by forcing a stream of air through it by bellows. Two of the small pipes had valves placed in them, one loaded with a weight equal to thirty-five times that of the atmosphere, and the other loaded to thirty-seven atmospheres; the boiler was quite filled with water, and was heated to between 300 and 400 degrees of Fahrenheit. When it had acquired this temperature, which was ascertained by the common means of a gauge placed on the boiler, a small quantity of water was forced through the pipe, in which was the valve loaded to thirty-seven atmospheres: and as water is incompressible, an equal quantity was forced (by the difference of the pressures) through the small pipe, in which was placed the valve loaded with a weight equal to thirty-five atmospheres. This valve being raised, the water heated to between 300 and 400 degrees, was admitted into a horizontal chamber, and expanded into steam of great elasticity. In this horizontal chamber or cylinder, Mr Perkins placed a piston, and opened by the usual means a communication between both of its sides with the boiler and condenser; the expansion of the hot water into steam on one side of the piston, and the formation of a vacuum on the other, alternately gave a reciprocating motion. The steam was condensed under a pressure of five atmospheres, and Mr Perkins estimated the power of his apparatus at the difference between five and thirty-five atmospheres, or 430 pounds on each square inch of the piston.

Part of the water which was condensed at a temperature of 350 degrees was pumped back into the boiler, as in common engines, and thus saved so much of the heat which would otherwise have been wasted.

The safety of this apparatus is insured by the usual contrivances of a steelyard valve, loaded with that weight which the boiler has been proved to be able to sustain without danger of bursting. The experimental boiler Mr Perkins considered to be equal to resist an internal pressure of four thousand pounds on the inch, being eight times greater than any strain the safety valves would admit its being subjected to. As an additional security he also revived the mode of making a safety valve, by connecting with the boiler a pipe of metal of that strength which would resist a pressure of a certain intensity, and burst when that force exceeded a certain limit; and he tried the efficacy of this scheme, by increasing the pressure of the water until the pipe gave way. A good engraving of the apparatus, and a very minute detail, are given in the third and sixth numbers of that useful miscellany the *Mechanic's Magazine*, to which we refer those readers who may think our explanation requires graphic illustration.

We have stated that Mr Perkins' boiler contained eight gallons, and that it was three inches thick; his cylinder was two inches in diameter and eighteen inches long, and the piston made a stroke of twelve inches: this was estimated as equal in effect to a Watt-Engine of ten-horse power, with an expense of only two bushels of coals per diem. The machine and its appendages stood in an area of forty-eight feet; and with the exception of the cylinder and its piston, the apparatus was considered in strength and dimension as sufficient for an engine of three times the power.

Mr Perkins has suggested the application of his invention to the boilers of engines of every construction, by the mete removal or disuse of their furnaces. He attaches his patent boiler to any common boiler; and the water which is heated under a great pressure instead of being forced into the piston cylinder, is forced into the water which is contained in the common boiler, and heats it to any degree suited to the nature or trim of the engine. By this mode he calculates that as much steam, having a pressure of four pounds to the inch, will be produced from one bushel of coals, as from nine by the common method.

From our description it will be seen that, properly speaking, Mr Perkins has made no improvement on the *Steam Engine*; for his experimental model was the same in all its details with the Watt-Engine; neither is using steam of a prodigious elasticity at all a novelty in steam apparatus. The fact also of water being capable of having its temperature raised under pressure; had been long known; but the method of heating the water subjected to this pressure, and the simple and effective manner of producing and continuing it, may possibly yet rank among the most important inventions of the time.

What effect the adoption of Mr Perkins' boiler may have in diminishing the expenditure of fuel, remains yet to be decided by experiments on the great scale. A moiety of the saving announced as being made by this invention, would be of immense national benefit. Could a saving of even a fourth part of the fuel be achieved by its use, we should consider this ingenious man neither to have laboured in vain for his interest, nor for his reputation.

How far the use of steam of such high temperature, by adding to the portability of the apparatus, will facilitate its extension to purposes in which its use at present is either doubtful or inexpedient, is considered to be abundantly obvious. Judging from the rapid strides the Steam Engine has made during the last forty-years, to become a universal first mover, and from the experience which has arisen from that extension; we feel convinced that every invention which diminishes its size, without impairing its power, brings it a step nearer to the assistance, of the "world's great labourers,"—the husbandman and peasant, for whom as yet it performs but little. At present it is made occasionally to tread out the corn;—What honours await not that man who may yet direct its mighty power to plough, to sow, to harrow, and to reap?

Chronological List of Patents

For improvements on the Steam Engine, and for saving fuel by the construction of the fire-place and boiler.

1698
Thomas Savery, London
Raising water by the elasticity of steam—Forming vacuum by condensing steam, to raise water by pressure of atmosphere

1705
Thomas Newcomen, John Cawley, Dartmouth, and Thomas Savery, London
Condensing the steam introduced under a piston, and producing a reciprocating motion by attaching it to a lever

1736
Jonathan Hulls, London
Steam-boat

1759
James Brindley, Lancashire
Boiler

1766
John Blakey, London
Improvement on Savery.

1769
James Watt, Glasgow
Invention of the condenser—Use of oil and tallow instead of water—Enclosing cylinder—Piston-jacket—Moving piston by steam against a vacuum—Steam-wheel

1769
John Stewart, London
Rotative from rectilineal motion

1772
J. Chrysel, London
Construction of furnace

1778
Matthew Washbrough, Bristol
Rotative from rectilineal motion

1781
John Steed, Lancashire
Crank movement
Jonathan Hornblower, Penryn
Two cylinders

1782
James Watt, Birmingham
Expansive engine—Six contrivances for regulating motion—Double impulse engine—Two cylinders—Toothed rack and sector to piston rod and beam—Semi-rotative engine—Steam-wheel

1784
James Watt, Birmingham
Rotative engine—Three parallel motions—Portable steam engine, and machinery for moving wheel carriages—Mode of working hammers and stampers—Improved hand gear—Mode of opening valves

1785
James Watt, Birmingham
Furnace for consuming smoke

1789
Thomas Burgess, London
A rotative from vibrating motion

1790
Bramah and Dickinson
Rotative engine

1791
James Sadler, Oxford
Rotative engine

1793
Francis Thomson, London
Two cylinders

1794
Robert Street, London
"Inflammable vapour force, by means of liquid air and fire, for communicating motion"—Turpentine falling on a hot iron, the vapour being exploded raises a piston placed in a cylinder.

1796
Valentine Close, Hanley
Saving fuel
John Pepper, Newcastle
Saving fuel
John Strong, Bingham
Valves
Francis Lloyd, Woolstanton
Furnace
William Batley, Manchester
"Mode of Working."

1797
Edmond Cartwright, Middlesex
Condensing engine—Piston—Rotatory engine
John Grover, Chesham
Boiler and furnace

1798
Thomas Rowntree, London
Furnace and boiler
Jonathan Hornblower, Penryn
Rotative engine
William Rayley, Newbald, York
"Philosophical furnace and boiler, with an actuating wheel as an appendage."
George Blundel, London
Machine for saving fuel
John Dickson, Dock-head
"Method of constructing"
Francisco Rapozo, Lisbon
Cylinder and valves
G. Quieroz, London
Cylinder—Boiler

1798
Robert Delap, Banville
Economical boiler
John Wilkinson, Castlehead
Boiler—Saving fuel
Marquis de Chabannes
Improving fuel
Matthew Murray, Leeds
Boiler—Damper—horizontal cylinder
A. G. Eckhardt, London
Saving fuel
W. Murdock, Redruth
Cylinder—Valves—Rotative engine
James Burns, Glasgow
Saving fuel—Furnace
James Bishop, Connecticut
Rotative engine
Samuel Rehe, London
"Machine for transmitting force of any fluid."
Rev. T. Cooke, London
"Carbo frugalist—an effectual mode of applying fire to caldronic implements."

1800
Peter Devey, London
Improved fuel
Phineas Crowther, Newcastle
Crank motion
John and James Roberton, Glasgow
Furnace for consuming smoke—Application of steam

1801
Edmond Cartwright, Middlesex
Portable engine—Regulating velocity
Richard Wilcox, Bristol
Engine and Furnace
William Hase, Saxthorpe
Cylinder—Boiler
James Anderson, Mounie
Saving fuel
Matthew Murray, Leeds
Air pump—Packing stuffing—Valves—Parallel motion
Timothy Bramah, Pimlico
Valve
Earl of Stanhope, London
Saving fuel
George Medhurst, London
Circular into rectilinear motion
George Stratton, London
Saving fuel
James Glazebrook, Colebrooke Dale
"Working machines by means of properties of air."
Robert Young, Bath
Saving fuel
William Symington, Kinnaird
Engine for steam-boat—Rotatory motion without a lever or beam

1802
James Sharples, Bath
"Mechanical powers applicable to steam-engine."
Thomas Parkinson, London
Conveying fluids

Richard Trevithick and Alexander Vivian, Cornwall
High pressure engine
Bryan Higgins, Leeds
Portable engine
Matthew Murray, Leeds
Portable engine
Thomas Martin, Brentwood
"Applying fire by means of certain machinery to heating."
Thomas Saint, Bristol
Boiler and furnace
Joseph Lewis, Brimscombe
Furnace
Matthew Billingsley, London
Boring cylinders
Richard Wilcox, Bristol
Furnace—Boiler—Air pump

1803
John Leach, Merton Abbey
Improvements in boiler
Arthur Woolfe, London
Boiler of tubes
Edward Stephen, Dublin
Saving fuel
Bryan Donkin, Dartford
Rotatory engine
John Edwards, London
Saving fuel
William Freemantle, London
Cylinder—Valves—Parallel motion—Pump

1804
Richard Wilcox, Bristol
Furnace and boiler
James Barrett, Saffron Walden.
Saving fuel.
Arthur Woolfe, London
Two cylinders and high pressure steam boiler

1805
James Rider, Belfast
Cylinder—Regulator
Charles Coe, London
Application of heat
Jonathan Hornblower, Penryn
Steam wheel
William Earle, Liverpool
"Working and constructing."
John Stevens, London
Boiler
Alexander Brodie, London
Boiler and furnace
James McNaughton, London
Saving fuel
James Boaz, Glasgow
Improvement on Savery.
Arthur Woolfe, London
Cylinder—Piston
Ralph Dodd, London
Saving fuel
William Deverell, Blackwall
Furnace and boiler
Samuel Miller, London
Saving fuel
John Trotter, London
Steam wheel

1806
William Lester, London
"Rotatory motion or engine."
T. Bourne, W. Chambers, and C. Gould, Warwick
Roasting meat by power of steam
Ralph Dodd, London
Simplification of machinery
John Lamb, New York
Application of heat
R. Wilcox, London
Rotative engine

Josias Robbins, Liverpool.
Furnace
William Nicholson, London
Application of steam

1807
Allan Pollock, Glasgow
Saving fuel
Henry Maudsley, London
Portable engine
Ralph Dodd, London
Economy of heat
James Bradley, London
Iron for furnace bars

1808
Thomas Mead, Hull
Steam wheel
Thomas Price, Bilston
Application of steam
James Linaker, Portsmouth
Steam boat
Thomas Smith, Bilston
"Certain improvements on steam engine."
J. Cowden and J. Partridge, London
Saving of fuel
Thomas Preston, London
Construction of furnace

1809
James Grellier, Aldborough-Hatch
Saving of fuel
Mark Noble, Battersea
New invented steam engine
John Murray and Adam Anderson, Edinburgh
Application of heat
W. C. English, Twickenham
Saving fuel
Edward Lane, Stoke-on-Trent
Improved rotative engine

John F. Fesenmeyer, London
"Constructing and Working."
J. F. Archbold, London
Application of heat
William Johnson, Blackheath
Heating fluids
Nugent Booker, Lime Hill, Dublin
Saving fuel
Richard Scantlebury, Redruth
"Certain improvements."
Samuel Clegg, Manchester
Steam wheel

1810
David Cook, London
Heating fluids
Arthur Woolfe, London
Constructing—Working—Saving fuel
William Docksey, Bristol
Application of heat
William Clerk, Edinburgh
Regulation of heat
William Chapman, Newcastle
Steam wheel
Richard Witty, Hull
"Makings—Arranging—Combining."
John Justice, Dundee
Application of heat
Stedman Adam, Connecticut
"Certain improvements."
John Craigie, Quebeck
Saving fuel

1811
Joseph Miers, London
Saving fuel
Richard Witty, Hull
Additions to former patent
Charles Broderip, London
"Certain improvements in construction."

Michael Logan, Rotherhithe
"Generation of fire."
William Good, London
Valves
John Trotter, London
Improvements in application of steam
George Gilpin, Sheffield
Application of steam
Henry James, Birmingham
Steam boat
Thomas Deakin, London
Saving fuel

1812
John Sutherland, Liverpool
Boiler and evaporating vessels
Henry Osborn, Bordesley
Manufacture of cylinders
R. W. Fox and Joel Lean, Falmouth
Certain improvements—Additional apparatus
Henry Higginson, London
Steam boat
Jeremiah Steele, Liverpool
Application of heat
William Onions, Paulton
Steam wheel

1818
Robert Dunkin, Penzance.
Saving fuel.
William Brunton, Butterly
Constructing and erecting engines
John Barton, London
"Various improvements."
John Sutherland, Liverpool
Furnace
Joseph White, Leeds
Improvements on engines
Charles-Broderip, London
Boiler

1814
W. A. Noble, London
"Improved steam engine."
John Rastrick, Bridgenorth
"Certain improvements."
R. W. King, London
Boiling water
Thomas Tudal, York
Steam carriages
John Slater, Birmingham
Boiler
R. Dodd—G. Stephenson, Killingworth
Steam carriages

1815
William Lash, Northumberland
Furnace
H. Houldsworth, Glasgow
Discharging condensed water
Matthew Billingsley, Bradford
"Certain improvements."
Richard Trevithick, Cambron
Piston—Rotative engine
William Moult, London
Furnace
John Cutler, London
Supplying fuel
W. & M. Beavon, Glamorgan
Furnace
Marquis de Chabannes
Saving fuel

1816
J. T. Dawes, Bromwich
Parallel motion
G. F. Muntz, Birmingham
Destroying smoke—Saving products—Furnace
Bryan Donkin, Surrey
Boiling water.

Alexander Rogers, Halifax
Saving fuel—Setting boilers
Philip Taylor, Bromely
Applying beat
William Stenson, Coleford
"Improved engine."
Robert Stirling, Edinburgh
Saving fuel
George Bodley, Exeter
Certain improvements
Joseph Turner, Layton
Rotatory engine
John Neville, London
"New means of generating and applying steam."
Joseph Gregson, London
Supplying fuel
William Losh, Newcastle
Furnace

1817
W. A. Osborne, Bordesley
Boring cylinder
George Mainwaring, Lambeth
"Improvements in steam engine."
John Oldham, Dublin
Steam-boats
George Stratton, London.
Saving fuel.
Moses Poole, London
"Certain improvements."

1818
Lord Cochrane and A. Galloway
Machine for consuming smoke
William Moult, London.
"Certain improvements."
Alexander Haliburton, Wigan
Furnace
Philip Taylor, Bromely
Applying heat

John Munro, London, and Barnabas Langton, New York
"Certain improvements.'
Joshua Routledge, Bolton-le-moor
Rotatory engine
James Ikin, Christchurch
Furnace bars
William Church, London
Certain improvements
William Johnston, London
Furnace—Destroying smoke
Marquis de Chabannes, London
Boiler of tubes
Jones & Plimley, Birmingham
Certain improvements on engine
Henry Creighton, Glasgow
Regulating admission of steam
John Malam, London
"Certain improvements."
Sir W. Congreve, London
Steam wheel
James Frazer, London
Junction of tunnels in boiler
Richard Wright, London
Construction—Subsequent employment of steam
John Seaward, London
"Raising steam."
William Brunton, Birmingham
Furnace
George Killey, Briggen
"Improvements in the construction."
John Pontifex, London
Improvement on Savery.

1820
John Oldham, Dublin
Additions to former Patent
William Carter, Middlesex
"Certain improvements."

John Barton, London
Engines and boilers for steam boats
John Hague, London
Improvements in making and constructing
John Wakefield, Manchester
Furnace and mode of feeding fuel
William Brunton, Birmingham
Furnace
Joseph Parkes, Warwick
Consuming smoke
Job Rider, Belfast
Rotatory engine
John Moore, Dublin.
Rotatory engine.
William Pritchard, Leeds
Furnace

1821
William Aldersley, Middlesex
Certain improvements
John Bates, Bradford
Feeding furnace
Thomas Masterman, London
Steam wheel
Robert Delap, Belfast
Steam wheel
Robert Stein, London
Certain improvements
Jonathan Dickson, London
Transmitting heat
Henry Pennock, Penzance
Furnace
Peter Devy, London
Preparing fuel
Henry Brown, Derby
Furnace—Consuming smoke
Philips London, London
Furnace
Aaron Manby, Horsely
Manufacture of engines

Thomas Bennet, Bewdley
Certain improvements
Francis A. Egells, London
Improvements in details
Sir W. Congreve, London
Addition to former patent
Charles Broderip, London
"In the construction"
Neil Arnot, London
Furnace and boiler
Julius Griffith, London
Steam carriage

1822
Richard Ormrod, Manchester
Boiler
John Gladstone, Castle Douglas
Construction of steam vessels
Alexander Clark, Leuchars
Condenser and boiler
Jacob Perkins, London
Boiler—Heating water under a great pressure
John Bambridge, London—Amos Thayer, Albany, United States
"Improvements in rotatory engine."
Thomas Leach, London
Steam wheel
G. H. Palmer, London
Furnace and destroying smoke
George Stratton, London
Consuming smoke
G. Stephenson, Long Benton
"Certain improvements"
M. J. Brunel, London
"Certain improvements"
John Stanley, Manchester
Supplying fuel to furnace
T. Binns and J. Binns, London
Boiler

1823
Nathaniel Partridge, Bowbridge
Furnace
Wm. Johnson, Great Totham
Boiler and furnace
Thomas Neville, Surrey
Boiler and furnace
William Jessop, Butterly
Metallic piston
Sir Anthony Perrier, Edinburgh
Furnace and boiler
M. J. Brunel, London
"Certain improvements."
Jacob Perkins, London
Boiling and evaporating by steam
Thomas Peel, Manchester
Rotatory engine
Jacob Perkins, London
Improvements in boiler
James Smith, Droitwich
Boiler
J. Fisher and J. Horton, West Bromwich
Boiler
William Jeakes, London
Water regulator to boiler
Joseph Bower, Leeds
Omission of air pump
William Wigston, Derby
"Certain improvements."
Robert Higgin, Norwich
Destroying smoke
James Surrey, Battersea
Furnace
James Christie, London
Furnace

Notes

1. The *Spiritalia* was first edited by Commandine, in 1571. It is also printed in the splendid folio collection of the works of the *Ancient Mathematicians*, published at Paris in 1693. The Greek text is accompanied with a Latin translation. The descriptions of the two machines, we have described, are in page 202 of that edition.
2. It is a remarkable circumstance, too, that this Greek scheme should be revived as an improvement upon the almost perfect modern mechanism; first by Kempel, a German, about 1785; and by a Mr Sadler of Oxford in 1791, at which time he obtained a patent for his invention.
3. Young's *Catal. Nat. Philos.* p. 263.
4. In 1597.
5. *Les Raisons des Force mouvantes avec divers Desseins de Fontaines*, folio, Paris 1624. Dr Brewster gives the date of 1615 to De Caus's book; if correct, it is probably that of an earlier edition, which we have not seen.
6. Branca's account of his contrivance is contained in a folio volume of machines, which he dedicated in 1628 to a M. Canci, Governor of Loretto. It was published at Rome in 1629, under the title of "Le Machine diverse del Signior Giovanni Branca." Our engraving is contained in plate XXV of that collection.
7. Partington's *Historical Account of the Steam Engine*. London, 1822.
8. "New and useful Inventions for Water Works: a work both useful and delightful for all sorts of people; translated into English by John Leak." The plates appear to have been those used in the French edition.
9. *Mathematical Magic*, 1648.
10. "Edward Somerset," says Walpole, "appears in a very different light in his public character, and in that of an author. In the former, he was an active zealot; and in the latter, a fantastic mechanic: in both very credulous." He was sent by King Charles into Ireland to negotiate with the rebel Catholics; but he is said to have exceeded some instructions, and to have forged others; which were disowned by the King; though he shielded the Marquis (then Earl of Glamorgan) from the consequences. "With all his affection for the Earl, the King mentions, in two of his letters, his want of judgment: perhaps his Majesty was glad to trust to his indiscretion, with which the Earl seems greatly furnished. We find him taking oath upon oaths to the Pope's nuncio, with promises of unlimited obedience both to his Holiness and to his delegate; and begging five hundred pounds of the Irish clergy, to enable him to embark, and fetch fifty thousand pounds; like an alchemist who begs a trifle of money for the secret of making gold. In another letter he promises two hundred thousand crowns, ten thousand stand of arms for foot, two thousand cases of pistols, eight hundred barrels of powder, and thirty or forty ships well provided." (Walpole.) And all that, when, according to a contemporary, "he had not a groat in his purse, or as much gunpowder as would scare a corbie."

11. The manuscript of the *Century of Inventions* is preserved among the Harleian Papers in the British Museum, and is numbered 2428 in that collection.
12. The *Century of inventions* was first printed in 12 mo. in 1683. It was reprinted in 1746, and was then supposed to be edited by Desaguliers. Another edition was printed at Glasgow in 1767, after Mr Watt had invented his engine. A third reprint is dated London, 1786: a fourth was edited by Mr John Buddle, of Newcastle, in 1813; to this is annexed an historical account of the Steam Engine for raising water. The entire work is also to be found in the first volume of the Repertory of Arts; and in the second volume of the third edition of Gregory's Mechanics.
13. "He appears to have been a nobleman of much knowledge and ingenuity; but his descriptions, or accounts of his inventions, seem not so much intended to instruct the public, as to raise wonder; and his encomiums on their utility and importance are to a degree extravagant, resembling more the puff of an advertising tradesman, than the patriotic communications of a gentleman. He was indeed a projector." Robison, *Ency. Brit.* art. *Steam Engine.*

    Mr Millington's apology for the early neglect of Lord Worcester's schemes is by far the most ingenious but even he admits "that several of his contrivances appear so extravagant, and so far beyond the reach of human power, that many have doubted whether they were invented or not." *Epitome of Nat. Phil.* vol. I, 1823.
14. "A Treatise on Propelling Vessels by Steam, by Robertson Buchanan, Civil Engineer," Glasgow, 1816, p. 16.
15. Dr Brewster has made the correction, although in other words. He reads, "One vessel of water being consumed, another begins to force, and then to fill itself with cold water." This is not only at variance with the spirit of the text, but with the process,—for the water is *forced.* The latter part of this unintelligible description, in the original, ought probably to be read as if the Marquis's apparatus consisted of two vessels similar to De Caus's, *without a separate boiler.* It would then be clear enough, and the Doctor's emendation would follow as a matter of course.
16. *Edin. Encyclop. Memoir of Hooke* by Brewster.
17. Robison's *Mechanical Philosophy*, 1828. A reprint of the various articles which were first published by Dr Robison in the *Encyclopaedia Britannica*. The notes to the article Steam Engine, in this collection, were written by the late venerable Mr Watt of Soho.
18. Harleian MSS. No 5771.
19. In the last edition of *Ferguson's Mechanics* by Dr Brewster, the expansion of steam is stated as it was given by Ferguson after Desaguliers, at 14,000 times; in no part of the excellent supplementary volume is there a correction made, or the true expansion stated—an error of moment, as the book has deservedly found its way into the hands of almost every mechanic.
20. The *Acta Eruditorum* of Leipsig for 1685 contain some communications by Papin. One of these is a description of a new machine to raise water, which is further noticed in the same Journal in June, and again in the August following; and in the *Nouvelles de la République des Lettres*, for July 1687, is a reply by Papin, to some objections raised against this apparatus by M. Nuit. He had invented the mode of dissolving bones by Steam of a very high temperature in 1681; an account of which he published in English in that year, and with some improvements in 1652 in French, under the title of "Le Manière d'emollir les Os et de faire cuire toutes sortes des Viandes en peu de tems et à peu de fraix." 'In this "Digester" Papin first introduced a "*Safety-valve.*"
21. *Acta Eruditorum*, p. 410.
22. "Recueil des diverses Pièces touchant quelques nouvelles Machines:" *à Cassel*, 1695. Extract *Phil. Trans.* 1697, p. 483.
23. "It was not until *after* Savery *had obtained* his patent, that the attention of Papin *was at all directed* to the means of obtaining *a moving power by Steam*: for *all* his former investigations had been confined to the nature and temperature of *Steam when prevented from escaping.*" Millington, Epit. p. 255. Savery's patent was *dated* 1698! A few pages farther on, this author, however, acknowledges, "Papin *was engaged* in some projects for producing a moving power through the agency of atmospheric pressure, and transmitting it to great distances by pipes; though it is not evident that he had at all made up *his mind* as to the best means of producing the *necessary vacuum*; for at one time large air-pumps to be worked by a powerful mill were proposed, at another the firing of gunpowder, *and lastly the production and condensation of Steam.* Although *all* these schemes *were*

published, it does not appear that any one *took advantage of them*, or constructed an engine upon *such* principles, till the time of Newcomen!" p. 259.

24. Harris, in his *Lexicon Technicum*, art. *Engine*, 1704. The engraving in this dictionary is the original one which was prefixed to the *Miner's Friend*, in 1702.
25. *Miner's Friend*, 1702, p. 6.
26. *Experim. Philos.* vol. ii. p. 466.
27. Robison, *Encyc. Britan.* art. *Steam Engine*: see also his *Mechanical Philosophy*, vol. ii. p. 48.

    Switzer, who was personally known to Savery, gives a somewhat different version of the story. "No contrivance," he observes, "for raising water is more justly surprising than the Fire Engine, the particular contrivance, and sole invention of a gentleman with whom I had the honour long since to be well acquainted. I mean the ingenious Captain Savery, some time since deceased, but then a most noted engineer, and one of the commissioners of sick and wounded. This gentleman's thoughts were always employed in hydrostatics, or hydraulics, or in the improvement of water-works; and the first hint from which it is said he took this engine was from a tobacco-pipe, which he immersed to wash or cools it, as is sometimes done. He discovered by the rarefaction of the air in the tube, by the heat or steam of the water, and the gravitation of impulse of the exterior air, that the water was made to spring through the tube of the pipe in a wonderful, surprising manner; though others say, that the learned Marquis of Worcester, in his 'Century of Inventions,' (*which book I have not seen,*) gave the first hint for this raising water by fire: It was a considerable time before this ingenious person (Capt. Savery), who has been so great an honour to his country, could, as he himself tells us, bring this his design to perfection, on account of the awkwardness of the workmen, who were necessarily to be employed about the affair; and I have heard him say myself that the very first time he played it was in a porter's house at Lambeth, where, though it was a small engine, yet it forced its way through the roof and struck up the tiles, in a manner that surprised all the spectators." p. 324. "Introduction to a General System of Hydrostatics, &c. by Stephen Switzer," 1729.

28. "When the Marquis's loose and vague description is recollected, and that he does not descend into the minutiae of executive construction; there appears to be no strong reason for depriving the captain of the title of an inventor!" *Crit. Nat. Philos.* p. 258.
29. "In Lord Worcester's time the machine was not practically introduced, and it was soon forgotten. Savery's engines were constructed in a manner precisely similar, and it is uncertain whether he adopted the Marquis of Worcester's ideas, or re-invented a similar machine." Dr Thomas Young, *Nat. Phil.* Vol. 1, p. 356.
30. "How useful it is in gardens and fountain works may or might have been seen, in the garden of that right noble peer, the present Duke of Chandois, at his late house at Sion Hill, where the engine was placed under a delightful banqueting-house, and the water being forced up into a cistern on the top thereof, used to play a fountain contiguous thereto in a very delightful manner." *Switzer*, Vol. 2, p. 884.
31. In the *Miner's Friend*, in which he (Savery) describes the invention as his own, the propelling or rowing boats or vessels by paddle-wheels, as now practised in our steam-boats, is particularly noticed. The Steam Boat, therefore, which is generally considered as one of the most recent applications of the force of steam to useful purposes, it will be seen is by no means a modern invention, but is nearly coeval with that of the engine itself."—*Millington*. This inference is not, however, to be drawn from Savery's words, for he nowhere in his *Miner's Friend* suggests the application of his engine to move steam-boats—*but to pump out* the water from ships. After suggesting the scheme of raising water from a pond by his engine, and allowing it to fall on a water-wheel, which would give motion to a mill, he notices the vast variety of millwork, and says, "Had I leisure to comment thereon, and give you an account not only of the vast variety that I have seen and heard of, but (when encouraged) what may yet be brought to work, by a *steady stream*, and the *rotation* or circular action of a water-wheel; it would swell these papers to a large volume." p. 28. This refers to a revival, by Savery, of a method practised centuries before his time to propel ferry-boats by the rotation of a water-wheel. A figure, on a large scale, of this mechanism is given in the Second Volume of Harris's *Lexicon Technicum*, published in 1710. The water-wheels or paddles are placed on each side of the ship, as in the common construction of modern steam-

boats—but these wheels are *not* moved by steam. A trundle fixed on the axis of the paddle-wheels, works into a wheel placed on the drum-head of the capstan; the capstan bars being turned round in the usual manner, give motion to the paddles; a contrivance precisely similar to that described by the late Mr Robertson Buchanan, in his *Treatise on Steam-boats*, as being in use at a ferry near New York, but with the slight difference of the capstan bars being moved by animal instead of manual labour.

32. *Miner's Friend.*
33. "New Improvement of Planting and Gardening, both philosophical and practical," 1717.
34. Switzer's *System of Hydrostatics*.
35. "Water in its fall from any determinate height has simply a force answerable, and equal to the force which raises it; so that an engine which will raise as much water as two horses working together at one time in such a work can do, and for which there must be constantly kept ten or twelve horses for doing the same, then I say such an engine will do the work or labour of ten or twelve horses." *Miner's Friend*. This is the true measure of the power of an engine when compared with the power of horses. Modern engineers affix a lower value to the same unit of comparison; and they state an engine to have the power of fifty horses (say) when it is equal to the raising of as much water in eight hours as fifty horses could raise, working that number of hours daily. But as the engine can be continued in operation throughout the day of twenty-four hours, this would require one hundred and fifty horses, and by Savery's mode would be called an engine of one hundred and fifty horse power.
36. Memoires de 1'Académie des Sciences, année 1699; see also Leopold, *Theatrum Machinarum*. Leipsig, 1724. Tab. 53, figura 2.
37. Prony, *Nouvelle Architecture Hydraulique*, vol. II. p. 90.
38. Blakey.
39. *Nouvelle Archit. Hydraul.* Vol. II. p. 200.
40. *Traité d'Hydrodynamique*, p. 306.
41. Belidor admits with great candour, "that, although the Marquis of Worcester was (in his opinion) the first in England who mentioned, in intelligible terms, a machine for raising water by fire, in a small tract called the *Century of Inventions*; yet we cannot deny to Captain Savery to have been the first to execute these sort of machines in Great Britain: this is attested by many letters written to me on that occasion, by the gentlemen of the Royal Society. There is one in which mention is made of a Mr Newcomen having contributed very much to bring it to its present perfection. Another proof that this machine took birth in England, and that it excels every other that hath been tried in France or Germany, is, that all the Fire Engines that have been constructed abroad have been executed by Englishmen." *Arch. Hydraulique*, Tom. II. p. 300. Gensanne, the inventor of a self-acting mechanism to be attached to Savery's Engine, begins the description of his model by saying that his apparatus is an improvement on a machine "which every one knows was *invented by M. Savery*, and executed on a great scale at London and in other parts of England." *Machines Approuvés*, Tom. VII. p.280.
42. *Mech. Phil.* vol. II. p. 49.
43. We should not have noticed these very palpable mistakes, had they not been made by a writer whose work is deservedly of high authority on this subject; and more particularly as they have been allowed to remain without correction or comment in the late collection of his mechanical papers; which from its lower price and convenient form, may find its way into more hands than they did when forming part of the *Encyclopaedia Britannica*. They have already been the means of propagating some erroneous notions and statements among later writers. So late as 1817, Mr Farey, after having taken considerable pains to arrange the date of Papin's machines, thus speaks of his inventions:—"We have copied the figure of Papin's engine from Belidor, that our readers may be able to compare it with Captain Savery's, and judge of the authority upon which M. Bossut has said, that the first notion of the Steam Engine was certainly owing to Dr Papin, who had not only invented the digester, but had, in 1695, published a little performance, describing a machine for raising water, in which the pistons are moved by *the vapour of boiling water alternately* dilated and *condensed*. Now *the fact* is, that Papin's publication *was in* 1707, in which he concedes the invention to Savery. He had occasionally, *before that*, published several inventions in the 'Acta Eruditorum,' *in which cylinders*

and pistons were to be employed, but they were not intended to be worked by steam, (rare candour,) but by gunpowder and air"!! Rees' *Cyclopedia,* art. *Steam Engine.* Mr Farey not having referred to the *Encyclopedia Britannica* as his authority, nor marked the passage as a quotation, it must be taken as conveying his own opinion in his own words. His additions to the Doctor's statement make it still more inconsistent. Does Mr Farey dispute the fact as stated by Bossut, and call the machine of 1695, an *Air Engine*? It is a quibble to say Papin's book on *the* Steam Engine was published in 1707. Papin's account of *a* Steam apparatus was dated in 1707; but is this the one alluded to by Bossut? Neither could Papin concede the invention to Savery, for he gives it to the Elector of Hesse. Throughout his book, Papin nowhere says that his engine is the same in principle or operation with Savery's; but claims merit for a totally different apparatus, and justly: excepting the mode of impelling water into the air vessel by the elasticity of steam, (a contrivance which belongs neither to him nor to Savery,) there is no feature of resemblance between their engines. In the same article, we have the "*first Steam Engine* with a piston, made by Papin 1707." In another part, "Papin's engine is far inferior to Savery's, and is only a return to the Marquis of Worcester's idea: "a position which may be admitted, when Mr Farey *constructs an Engine according to the Marquis's idea.*

44. *Nat. Phil.* p. 467.
45. "In the possession of the Royal Society."—Robison.
46. "Savery, however, claims the invention as his own; but Switzer, who was personally acquainted with both, is positive that Newcomen was the inventor. By his principles (as a quaker) being averse to contention, he was contented to share the honour and profits with Savery."—*Mech. Phil.* vol. II. p. 58. Savery *only* claimed the *method* of forming a vacuum by the condensation of steam—an essential part of Newcomer's apparatus; not, as might be inferred from the Doctor's observation, the invention of the engine. Robison calls Newcomen and Cawley "quakers;" Desaguliers, who is the better authority, says they were "baptists:" an error of trivial consequence, had not their sectarian principles been given by the Doctor as a reason for their conceding to Savery what did not belong to him.
47. Switzer's *Hydrostatics.*
48. "The French authors," according. to Mr Farey, "have claimed this engine also as the invention of their countryman Papin, *but without any reason.* Papin had gained a knowledge of the expansive force of steam in his digester, and he *invented* the mode of *working the pistons and cylinders by a vacuum* and the *pressure of the atmosphere:*" (this appears an unanswerable argument for Papin's claim of having invented the Atmospheric Engine:) "but he was not the *first* inventor of either of these, *Otto Guericke* and the Marquis of Worcester having discovered the same things long before him!!" When the German moistened the valves of his rude machine, it is probable that the water he employed was raised by a "sucking" pump: therefore, according to Mr Farey's logic, he was not the inventor of the air-pump; but an unknown ancient who contrived the sucking-pump, must be considered as having invented the air-pump, and the Atmospheric, or Papin's Engine. This, however, we should call a "sucking" inference, so refined as to be exquisitely foolish. It is certainly the first time, too, that the venerable Otto Guericke has been honoured as the author of an invention claimed for Papin, of forming a *vacuum* by the *condensation* of steam; and which Mr Farey, in the same paragraph, also assigns to Savery! "And farther, he (Papin) had no pretensions to claim *Savery's discovery* of the *condensation* of steam, upon which the engine of Newcomen depends." Savery made no *discovery*, but a novel, refined, and ingenious application of a well-known physical law to an important practical purpose.
49. Desaguliers. "It does not appear that the Marquis of Worcester knew any thing of the use of an injection, as the machine described by him operated only by the expansive force of the steam; whereas the injection was used in Savery's engine from the beginning, and is in all probability his invention." Note by Mr Watt in Robison's *Mech. Phil.* vol. II p. 50. The venerable improver of the Steam Engine may have been writing from recollection when he stated this opinion. Condensation, in all Savery's engines, was produced by affusion of cold water on the outside of the receivers. We have seen that in an engine erected in 1710 by Savery himself, the injection was not used; and in the *Miner's Friend* there is no mention of any contrivance like it.
50. "To scog, to be found scogging, to be a scogger, are terms in very common use in the north of Yorkshire, and convey exactly the meaning of the terms to skulk, to be found skulking, to be a

51. Desaguliers' *Nat. Phil.*
52. Tallow was used in these Engines to lessen the friction, but not to keep them air-tight.
53. The origin of packing the piston is thus given by Desaguliers:—"Having screwed a large broad piece of leather to the piston which turned up the sides of the cylinder two or three inches, in working it wore through and cut that piece from the other, which, falling flat on the piston, wrought with its edge to the cylinder, and, having been in a *long time*, was worn very narrow; which being taken out, they had the happy discovery, whereby they found that a bridle-rein, or even a soft thick piece of rope, going round, would make the piston air and watertight."— Desaguliers' *Nat. Phil.* Hornblower observes, "We need not say any thing to the practical engineer about *leathering* a *steam piston*. Nor is it necessary to comment on the Doctor's acquaintance with steam and leather in contact."—Gregory, *Mech.* vol. ii. p. 358. 1st edition.
54. *Nat. Phil.*
55. Desaguliers' *Nat. Phil.*
56. Des. *Nat. Phil.*
57. *Theatrum Machinarum*, vol. ii. Tabula. 30.
58. *Machines approuvées*, p. 209. Tom. VII.
59. A Description and draught of a new-invented machine for carrying vessels or ships out of or into any harbour, port, or river, against wind or tide, or in a calm. By Jonathan Hulls. London, printed for the author, 1737, price 6d.
60. In page 41, Savery's scheme of moving ships by paddlewheels is described. Dr Brewster considers Hull's adaptation not of much moment: "The substitution of the power of horses, or of steam, or of heated air, in place of the strength of men, appears to us no invention at all: if it were, we should have numerous rivals contending for the honour of applying the Steam Engine to the threshing machine. When Mr Jonathan Hulls therefore, in the year 1786, took out a patent for the application of one of *Newcomen's* Steam Engines to a vessel for towing ships in and out of harbour, he merely proposed to substitute the power of steam in place of the power of men. His proposal was neither characterised by sagacity nor inventive genius; and the intermediate mechanism by which the reciprocating motion of the piston was converted into the rotatory motion of the paddle-wheel, or fans, as he called them, was clumsy and imperfect."—Ferg. *Mechanics*, vol. ii. p. 113. In anticipating a similar objection, Hulls gives a fair answer to it:—"If it should be said that this is not a new invention, because I make use of the same power to drive my machine that others have made use of to drive theirs for other purposes; I answer,—The application of this power is no more than the application of any common or known instrument used in mechanism for new-invented purposes."—Hulls, as quoted by Brewster. It is not always a fair way to judge of the value of a contrivance by its importance as estimated in times of comparatively refined invention. At this moment we should call the application of the Steam Engine to move Balloons, a very fine invention; although the Engine itself should be the identical one that had moved a threshing- machine or a coal-wagon.
61. *Machines Approuvées*, p. 300. Tom. VII.
62. *Architecture Hydraulique*, Tom. II. p. 300.
63. Smeaton, *Philosophical Transactions*, page 437. Vol. XLVII. 1752.
64. *Phil. Transactions*, 1741.
65. Hornblower, in *Greg. Mech.* p. 306. vol. II.
66. Hornblower, in *Greg. Mech.* p. 363, vol. II.
67. *Blakey sur les Pumpes à Feu.*
68. Robison.
"In 1781 the Abbé Arnal, Canon of Alais in Languedoc, entertained a thought of the same kind, and proposed it for working lighters in the inland navigation; a scheme which has been successfully practised (we are told) in America. His brother, a major of Engineers in the Austrian service, has carried the thing much farther, and applied it to manufactures; and the Aube Chamber of Mines at Vienna has patronised the project., (See *Journal Encyclopédique*, 1781.) But these schemes are long posterior to Mr Fitzgerald's, and are even later than the erection of several

machines driven by Steam Engines, which have been erected by Messrs Boulton and Watt. We think it our duty to state these particulars, because it is very usual for our neighbours on the Continent to assume the credit of British inventions"—Dr Robison, *Encyc. Brit.*

69. Dr Robison states that Mr Watt was the first who practised this mode of surrounding the fire-place with water. "The mode of conveying flame through water," says Mr Watt in a note to the Doctor's statement, "had been practised by others before my time, and was common in the Cornish mines. The inventor is unknown, but a person of the name of Swaine was a great propagator of the practice." Rob. *Mech. Phil.* art. *Steam Engine*. The patent in the text appears to have been unknown to Mr Watt; but in fact Brindley only revived the method. Sir Robert Moray and Dr Goddard, in the beginning of 1663, proposed "brewing beer in a kettle, having only a brass bottom; and in the middle thereof a globe of brass open at the lower end, into which the fire goes, whereby the brass of the rest of the kettle is saved." Glauber used *wooden casks* for boilers, and he boiled the water they contained by "a small copper globe, (which was attached to them) placed in a furnace heated with sea-coal;" "a contrivance," says Hooke, "which if prosecuted might perhaps be very beneficial to brewers, dyers, and such other trades as have occasion to make use of great quantities of water heated."

70. Historical Account of the Steam Engine.

71. It had, however, been used *indirectly* as a first mover of rotative machines, by being made to pump water to a given height, which was then conducted on a water-wheel of the common construction: a mode which was also sometimes resorted to where Savery's Engine was in operation.

72. Mr James Watt was born at Greenock in 1756. His father was one of the baillies (or magistrates) of the town; and in his narrow sphere was esteemed as a man of great benevolence. His grandfather and uncle were both respectable mathematicians: his uncle was author of a Survey of the River Clyde. From his infancy Mr Watt's constitution was of the most delicate kind, and to this may be attributed those retiring and studious habits, which were so remarkable in Mr Watt during the whole period of his long and distinguished life. After going through the usual course of education at the public Grammar School, at the age of sixteen, he was apprenticed to a mathematical instrument maker. Here he acquired habits of despatch and order in business; but by sitting in winter near the door of the workshop, he caught a severe cold, the effect of which he felt at times until he had attained the age of sixty years, when the severe and distressing head-aches it occasioned ceased to afflict him. At the end of a year, finding his health declining, he returned to Scotland, and settling at Glasgow, he began business on his own account. In the same year (1757,) he was appointed mathematical instrument maker to the university; and apartments were also given to him in the college, in which he lived and transacted his business.—*Memoir* by *Playfair*, in *Month. Mag.* for 1819.

73. Narrative of his invention in Robison's *Mech. Phil.* vol. II. art. *Steam Engine*.

74. This contrivance differs nothing from our Figure of Leupold's High-pressure Engine; and adds another instance, to the many already in existence, of similar inventions being made by different individuals, unknown to each other.

75. This employment was very different from that which now goes under the same name in London, and in the larger provincial capitals. It included not only the making and repairing the instruments used in the experiments in mechanics and natural philosophy, and those used in land-surveying; but also the manufacture, in a rough way, of all kinds of musical instruments; drawing lines on dials; cleaning and repairing wooden clocks; making and repairing fishing-tackle; and acting as a general sort of rough cutler. As an instrument-maker, Mr Watt was called on for his assistance in all these; but the greater part of his time was taken up in making and repairing drawing-pens, compasses, and plotting scales, and making balances and weights.

76. "Mr Watt's first attempt at the improvement of the Steam Engine, was to employ a wooden cylinder, which would transmit the heat more slowly: this had some effect, but did not answer in other respects; he was obliged to abandon it, as well as Mr Brindley, who had before tried the same thing."—Farey in Rees' *Cyclop*. Mr Brindley's trial was made on a very different part; he cased his *boiler*, or rather built it with wood, —not his cylinder.

77. "Mr Watt examined the hot water which issued from the eduction-pipe of several of Newcomen's Engines, and found it to vary from 142° to 174°, according to the load and other circumstances of the engine. He thought this might be taken as a fair indication of the internal heat of their cylinders."

78. Dr Tire, in his excellent *Dictionary of Chemistry*, gives the following interesting account of Mr Watt's original experiments on the latent heat of Steam. "In some conversations with which this great ornament and benefactor of his country honoured me a short period before his death, he described with delightful naiveté the simple but decisive experiments by which he discovered the latent heat of Steam. His means and leisure not then permitting an expensive and complex apparatus, he used apothecaries' phials: with these he ascertained the two main facts,—first, that a cubic inch of water would form about a cubic foot of ordinary steam, or 1728 inches; and that the condensation of that quantity of steam would heat six cubic inches of water, from the atmospheric temperature to the boiling point. Hence he saw that six times the difference of temperature, or fully 800 degrees of heat, had been employed in giving elasticity to steam; and which must be all subtracted before a complete vacuum could be obtained under the piston of a Steam Engine." Art. *Caloric*.

79. "Watt's account of his invention, Robison's *Mech. Phil.* vol. II, p. 117.

80. "Hornblower in his history of the Steam Engine, says "About the time that Mr Watt was engaged in bringing forward the improvement of the Engine, it occurred to Mr Gainsborough, the pastor of a dissenting congregation at Henley-upon-Thames, and brother to the painter of that name, that it would be a great improvement to condense the steam in a vessel distinct from the cylinder, where the vacuum was formed; and he undertook a set of experiments to apply the principle he had established; which he did by placing a small vessel by the side of the cylinder, which was to receive just so much steam from the boiler, as would discharge the air and condensing water in the same manner as was the practice from the cylinder itself in the Newcomenian method; that is, by the snifting valve and sinking pipe. In this manner he used no more steam than was just necessary for that particular purpose, which, at the instant of discharging, was entirely uncommunicated with the main cylinder; so that the cylinder was kept constantly as hot as the steam could make it. Whether he clothed the cylinder as Mr Watt does, is uncertain: but his model succeeded so well, as to induce some of the Cornish adventurers to send their engineers to examine it; and their report was so favourable as to induce an intention of adopting it. This, however, was soon after Mr Watt had his Act of Parliament passed for the extension of his term; and he had about the same time made proposals to the Cornish gentlemen to send his Engine into that country. This necessarily brought on a competition, in which Mr Watt succeeded: but it was asserted by Mr Gainsborough, that the mode of condensing out of the cylinder was communicated to Mr Watt by the officious folly of an acquaintance, who was fully informed of what Mr Gainsborough had in hand. This circumstance, as here related, receives some confirmation by a declaration of Mr Gainsborough the painter to Mr T. More, late secretary to the Society for the encouragement of the Arts, who gave the writer of this article the information; and it is well known that Mr Gainsborough opposed the petition to the House of Commons, through the interest of General Conway." Hornblower in *Gregory's Mech.* p. 361. vol. ii. 1st edition. On this extraordinary and disingenuous statement Dr Brewster gives the following comment:—"We believe and hope, for the sake of the memory of a very respectable man, that the conversation is not accurately represented. It remains upon record, that Mr T. More was examined as a witness on the trial of the cause Bolton *versus* Bull, in 1792, at which time Mr Hornblower himself was also examined as a witness, but on the opposite side from Mr More. Mr More on this occasion was asked, "whether he had read the specification of Mr Watt's invention, and whether, in his opinion it contained a disclosure of the principles of the Steam Engine?" To this question he answered, "I am fully of opinion that it contains the principles, entirely, clearly and demonstratively." He was then asked, "Did you ever meet with the application of these principles before you knew of Mr Watt's Engine?" His answer was, "I do declare I never saw the principles laid down in Mr Watt's specification, either applied to the Steam Engine previous to his taking it up, or ever read of any such thing whatever." It is not easy to reconcile these two answers given by that gentleman upon oath, with the words that Mr Hornblower has put into his mouth, p. 328. Mr Gainsborough's idea, whatever it was, was posterior by more than twenty years." *Edinburgh Review*, 1809.

Dr Robison, in describing this improvement of the condenser, speaks of it as if applied to Newcomen's Engine. This, however, does not appear ever to have been contemplated by Mr Watt: in a note he says, " From the first I proposed to act upon the piston with steam instead of the atmosphere, and my model was so constructed."

81. Mr Watt's narrative. Robison, *Mech. Phil.* p. 119.

This experiment has been differently related by some authors:—"The thought struck him (Mr Watt) to attempt the condensation in another place. His first experiment was made in the simplest manner. A globular vessel communicated by means of a long pipe, of one inch diameter, with the bottom of his little cylinder of four inches diameter, and thirty inches long: the pipe had a stop-cock, and the globe was immersed in a vessel of cold water. When the piston was at the top, and the cylinder filled with strong steam, he turned the cock. It was scarcely turned, when the sides of his cylinder (only strong tin plate) were crushed together like an empty bladder: this surprised, and delighted him." Robison.

"Mr Watt had nearly concluded (after casing his cylinder with wood) that the Steam Engine was incapable of further improvement, and was in as perfect a state as possible, when the happy thought occurred to him of condensing the steam in a separate vessel, instead of within the cylinder of the engine. The expedient was immediately tried in a very simple manner, as follows:—A globular copper vessel was made to communicate, by means of a pipe containing a cock to shut it, with a hot low tin cylinder, thirty inches long and four inches diameter. The globe was filled with steam by connection with a small boiler, while the tin cylinder was immersed in cold water; and upon opening the cock to form the communication between the two vessels, the sides of the tin cylinder were instantly crushed together, on account of its not being sufficiently strong to resist the pressure of the atmosphere,—so perfect was the vacuum that was formed within it. This *experiment exceeded his* (Mr Watt's) *most sanguine wishes*, and upon it *he founded all the most* important improvements which he was afterwards enabled to make in this important machine." Millington.

It is almost a pity to crush the interesting "little cylinder and globular vessel" a second time. But Mr Watt himself has thrown cold water on them; for he assures us, "that the globular vessel *only existed in his mind, and was never executed*:—the *tin* cylinder is a mistake; there never was any used in this model, the cylinder being of *brass*." *Mech. Phil.* vol. II. p. 109.

82. *Edinburgh Review*, 1809.
83. *Smeaton's Reports*, vol. I.
84. Patent dated 1769.
85. *Edinburgh Review*, 1809.
86. Mr Farey, art. *Steam Engine*, Rees' *Cyclopedia*:

"Dr Robison, (says Mr Watt in a letter to his friend Dr Brewster,) in the article Steam Engine, after passing an encomium upon me, dictated by the partiality of friendship, qualifies me as 'the pupil and intimate friend of Dr Black;' a description which not being then accompanied with any inference, did not particularly strike me at the time of its first perusal. He afterwards, in the dedication to me of Black's Lectures on Chemistry, goes the length of supposing me to have professed to owe my improvements upon the Steam Engine to the instruction and information I had received from that gentleman, which certainly was a misapprehension: as, though I have always felt and acknowledged my obligations to him for the information I had received from his conversation, and particularly for the knowledge of latent heat, I never did nor could consider my improvements as originating in those communications. He is also mistaken in his assertion in the preface to the above work, that I had attended two courses of the Doctor's lectures; for, unfortunately for me, the necessary avocations of my business prevented me from attending his, or any other lectures at college. In further noticing Dr Black's opinion that 'his fortunate observation of what happens in the formation and condensation of elastic vapour, has contributed in no inconsiderable degree to the public good, by *suggesting* to my friend Mr Watt of Birmingham; then of Glasgow, his improvements on the Steam Engine;' it is very painful for me to controvert any assertion or opinion of my revered friend; yet in the present case I find it necessary to say that he appears to me to have fallen into an error. These improvements proceeded upon the old established fact, that steam was condensed by the contact of cold bodies, and the later known one, that water boiled heats below 100; and consequently that a vacuum could not be obtained, unless the cylinder and its contents were cooled every stroke to below that heat."

Dr Thomson in his memoir of Dr. Black has followed Dr Robison. Dr Black's discovery of latent heat "constitutes the whole doctrine of heat as at present taught by the chemists, and which has been attended with more benefit to the world than any other discovery made during the eighteenth century, since it occasioned the improvements in the Steam Engine by Mr Watt; an instrument which has operated a complete change in our manufactures." In another paragraph the theory of latent heat is said to have "led to Mr Watt's improvement in the Steam Engine, which has produced such mighty changes in our style of mining and in our manufactures, and which has so enormously increased the power and resources of man."—*Edinburgh Encyclopaedia*, *Life of Black*.

87. Young's *Nat. Phil.* vol. I. p. 366.
88. "A Steam Wheel moved by the force of steam acting in a circular channel against a valve on one side, and against a column of mercury or other fluid metal on the other side, was executed at Soho upon a scale of six feet and tried repeatedly, but was given up, as several objections were found against it." Mr Watt in *Robison*, vol. II. p. 133.
89. Specification of Patent, 1796.
90. "One of the first engineers this country ever saw, erected one of those machines, the lever of which was so large as would require a cylinder equal to the power of near forty horses to give it motion."—Hornblower.
91. *Edin. Rev.* for 1809.
92. Memoir by Playfair. *Monthly Mag.* 1819.
93. When the "princely Bolton" had closed his long and active life, with a generous and grateful recollection of his virtues, Mr Watt thus speaks of his obligations to his lamented friend. "At the procuring of this Act of Parliament I commenced a partnership with Mr Bolton, which terminated with the exclusive privilege in 1800, when I retired from business; but our friendship continued undiminished to the close of his life. As a memorial due to that friendship, I avail myself of this, probably a last public opportunity of stating, that to his friendly encouragement, to his partiality for scientific improvements, and his ready application of them to the processes of art, to his intimate knowledge of business and manufactures, and to his extended views and liberal spirit of enterprise, must in a great measure be ascribed whatever success may have attended my exertions."
94. Mr Farey in *Rees Cyclo.* art. *Steam Engine*.
95. "We believe that until Mr Watt went into Cornwall, the *blowing valve* (in the condenser) had never been applied to any of his engines, it being the usual method to pump out the air by a temporary brake attached to the discharging pump; and that this valve was first applied by Mr Hornblower at an engine on a mine called Ting Tong, which engine was erected by him for the proprietors of the work." Hornblower in *Gregory's Mechanics*, p. 372, vol. II.
96. Watt's narrative.
97. Watt's narrative.
98. Farey in *Rees' Cyclopedia*, art. *Steam Engine*.
99. Hornblower.
    In the drawing presented to the House of Commons in 1772 by Mr Watt, a fly is placed on the axis of the sun-wheel.
100. Farey in *Rees' Cyclopedia*, art. *Steam Engine*.
101. "Mr Watt was too inoffensive a man to attack Prony; and when the injustice done was mentioned to him in 1810 or 1811, when he was in London, he said that it was true, but that he had seen de Prony, who had made a sort of an apology, or entered into an explanation. Mr Watt did not appear to wish to enter on the subject." Playfair's *Memoir*. Nothing more strongly marks the equanimity and greatness of Mr Watt's mind, than his indifference to the attempts of those who would detract from his claim as an inventor. Prony, however, never had either the candour or the grace to make an apology through the same medium that he attempted to do the injury.
102. "Prony's work," says Mr Farey, in *Rees' Cyclop.* "is little more than a description of the plates. These engines are not the best specimens of Mr Watt's invention; they were *all constructed in France by M. Perrier of Paris*, who, in 1780, erected a large engine at Chaillot, to pump up the water of the Seine for the supply of the town, and another of smaller dimensions on the opposite side of the river at Gros Caillou. These engines are still (1817) at work, and I visited them in 1814. They are upon the plan of Mr Watt's first Engines, though, for want of some attention to minute particulars, they do not produce any great effects. M. Perrier had visited England to obtain the

requisite instructions for making these engines." "Frenchmen," says Playfair, "are at great pains to conceal the origin and country of the Chaillot Engine;" they have even succeeded in concealing it from Mr Farey.

103. Hornblower erected several Engines on this scheme, with various contrivances to evade Mr Watt's patent for the condenser and air-pump; but into practice they were not found equal the original. And the proprietors of those engines, rather than run the chance of a decision by a Jury, submitted to pay to Bolton and Watt the sum these gentlemen commonly charged for the privilege of using their Engines. Mr Hornblower's preposterous application to Parliament in 1792, to extend the term of his patent, it need hardly be added, was refused; and *none have been erected since the patent has expired*. The valves, however, introduced by him were very ingenious and effective, and his mode of trussing the working-beam was highly creditable to his mechanical attainments.

104. One of his first trials was uncommonly ingenious. It consisted of a drum turning air-tight within another, with cavities so disposed, that there was a constant and great pressure urging it in one direction; but no packing of the common kind could preserve it air-tight with sufficient freedom of motion. He succeeded by immersing it in mercury, or in an amalgam, which remained fluid at the heat of boiling water, but the continual action of the heat and steam, together with the friction, soon oxydated the fluid, and rendered it useless. He then tried Parent or Harker's Mill, enclosing the arms in a metal drum which was immersed in cold water. The steam rushed rapidly along the pipe which was the axis, and it was hoped that a great re-action would have been exerted at the ends of the arms, but it was almost nothing. It was then tried in a drum kept boiling hot, but the impulse was very small, in comparison with the expense of the steam," Farey in *Rees' Cyclop*. art. *Steam Engine*, 1817.

105. In the *History of Steam Navigation* we ought not to omit the name of a M. J. A. Genevois, a clergyman in the Canton of Berne, who, in 1759, published at Geneva a book containing what he called a discovery of the "Great Principle." This was to concentrate power, by whatever means produced, into a series of springs, which might be applied to a variety of uses at the most convenient time, or in the most convenient manner afterwards. He suggested the application of his "great principle," to the mode of propelling a vessel by oars worked with springs. He also proposed the use of an Atmospheric Steam Engine, to bend the springs which were to move the oars, and also to work a "winged cart" when the wind failed, and a "winged machine" in any wind—even a quite contrary one. His favourite project, however, appears to have been to use the expansive force of gunpowder, to bend the springs of his oars. He came to England in 1760, to submit his book and plans to the Lords of the Admiralty, who desired him to extract and submit to them that part of his book relating to navigation.—This memoir he printed, with a plate containing figures of the mode of forming and using the oars and the "gunpowder cylinder."—We quote the following anecdote from his pamphlet. "It is true, an honourable gentleman, one of the members of the Navy Office, told me when I appeared before them on the 4th August 1760, that about thirty years ago, a Scotchman proposed to make a ship sail with gunpowder, but having found by the experiments made for that purpose, that thirty barrels of gunpowder had scarce forwarded the ship the space of ten miles, this invention had been rejected. To this I answered, that he acquainted me with a thing quite new to me; that his scheme was deservedly rejected, but that my work was of another kind. I have been since told that it was from the power of retrogradation of one or more cannons on the poop, this man had conceived the hope of forwarding the ship. This put me in mind of a trial a celebrated gentleman made many years ago on the Rhine, by the effusion of the water from a tub on the stern by a hole towards the prow. This was only a sport as for the Scotchman's work, it has nothing to do with mine but the thought of gunpowder." p. 20.

106. "Journal des Débats." Partington, *Historical Account of the Steam Engine*.
107. Brewster.
108. Dr Brewster, in his supplementary volume to Ferguson's Mechanics, describes the boat as double and having *wheels* in the centre. We have not been able to see the book. Mr Buchanan, in his treatise on Steam Boats, says the experiment did not succeed to Mr Miller's satisfaction.
109. "I have also reason to believe that the power of the Steam Engine may be applied to work the wheels, so as to give them a quicker motion, and consequently to increase that of the ship."—Mr Miller, as quoted by Dr Brewster.

110. Trans. *Roy. Irish Academy*, 1787.
111. *Annales de Chemie*, 1809.
112. Specification of patent, 1791.—*Repertory of Arts*, vol. III, First Series.
113. *Memoirs of Phil. Soc. of Lausanne* 1791; also p. 203, vol. IV of *Repertory of Arts*, First Series
114. *Handbuch der Mechanik*. Altenburgh, 1794.
115. The First Volume of the *Philosophical Magazine* contains an admirable engraving of this Steam Engine.
116. *Repertory of Arts*, vol. X. p. 7.
117. *Journal of Royal Institution*, Young's *Nat. Phil.* vol. II.
118. Buchanan on *Steam Boats*, page 7.
119. *Repertory of Arts*, p. 11, vol. XIII
120. *Repertory of Arts*, vol. XV. First Series.
121. *Transactions of American Philosophical Society*, vol. I. p.209.
122. The following curious speculation is given in the *Epitome of Nat. Phil.* page 322:

    "The construction of the high-pressure Engine is so nearly allied to those of Savery and Papin, that it requires but a very *small* extent of inventive genius to convert either of these, and particularly the last with a cylindrical steam vessel, into such a machine—for we have only to *suppose the circular piece of wood, already mentioned as floating on the water in that vessel, to be a steam-tight piston, equipped with a piston rod passing to the outside of the vessel, and the engine will be formed! for the force that was then described to be acting to depress the water, will now depress that piston*, the power of which may be transferred to any other purpose." This is very plausible; but Playfair has shrewdly observed in his memoir of Watt, that "any person who now sees with what facility the engines are managed, and the perfection with which they are made, might feel a *difficulty* in conceiving the difficulties that good workmen and men of genius found in managing the machine, and making it perform well even *after* it was invented—but this proves the truth of the old proverb, that practice makes perfect." What could be a more obvious step than the addition of a separate condensing vessel to prevent the cylinder from being cooled by the cold water thrown into it; yet this was the greatest and first of Mr Watt's improvements.

123. Unequal contraction and expansion are very often occasioned in the side of cast-iron boilers, when in action, by *cold water* falling on their surface. This might form a fracture, which, although not probably apparent at the moment, would yield to steam of an increased elasticity. Carefully covering the surface of the boiler by some body of a low conducting power, would not only be economical in point of heat, but expedient and necessary with regard to safety. We have witnessed on a small scale, this effect produced by the quick affusion of a small quantity of cold water, about 40°, on the side of a small boiler containing steam of about 230.
124. To a person complaining of the frequent derangement of his condensing Engine: Mr Watt once said, "Your Steam Engine, Sir, is like the horse whose work it is performing: the attention it requires is of the most ordinary kind, but it must have *some* attention."
125. See *Repertory of Arts for 1802*, and *Philosophical Magazine* for the same year.
126. Hornblower, in *Greg. Mechanics*, vol. II. p. 387.

    "Is there not some ground to fear that in this contrivance, besides the force lost by the action of the steam upon the edges of the vanes, there will be a considerable loss arising from the greater friction attending its operations than those of a common Steam Engine? In this steam-wheel there will be a great quantity of rough surface (that of the stuffing) exposed to frequent contact, and consequent resistance to the moving from the fixed parts. Besides, as the stuffed parts are here of great extent with regard to the magnitude of the machinery, and exhibit rapid variations of shape, they may, when brought into constant work, be found difficult to keep in order." Note by Dr Gregory.—The specification is printed at length in the *Repertory of Arts*.

127. *Historical Account of Steam Engine*, p. 46.
128. Trinidad is stated in this account to be the first West India Colony, in which the Steam Engine was introduced; but we believe that in Jamaica they were in action many years before this period.
129. *Repertory of Arts*, vol. VII. p. 322. Second Series.
130. 1806.
131. *Repertory of Arts*, vol. XVII. p. 130. Second Series.

132. *Idem*, vol. XX. p. 258.
133. *Buchanan on Steam-boats*, p, 7.
134. Observations on his Engine. *Repert. of Arts*, Vol. XV. p. 325. Second Series.
135. *Repertory of Arts*, Vol. XXXVII. p. 260. Second Series.
136. *Buchanan on Steam Boats*, p. 7.
137. Mr William Bull, of Chasewater, Cornwall, who was one of the three Cornishmen that accompanied M. Uvillé: Thomas Trevarthen, of Crowan, and Henry Vivian, of Camborne, were the other two.
138. *Geological Transactions of Cornwall*, vol. I, page 222.
139. Of a man whose discoveries and inventions reflect so much honour on the age in which he lived, and have produced such wonderful effects in the condition of the country, it may be considered as interesting to learn something of his attainments as a mechanic, and habits as a man of business:—"His mind has been compared to an encyclopaedia, that, open it at which part you will, some fact was stated, some truth illustrated, or some analogy traced" with the energy and comprehension of a mastermind. His knowledge was, however, of the useful kind, and, unlike many of his plodding and common-place contemporaries, he valued knowledge for itself not, for the manner in which it was imparted. He had no relish for what was called abstruse mechanical discussion; and on theoretic deduction he never placed, any dependence. In this he resembled Smeaton; every thing was done by a sort of feeling or tact, as if the knowledge had been born with him. In his opinions he never went either before or beyond the direct inference which could be drawn from an experiment; but so great was his sagacity, that few bearings of that experiment were omitted or overlooked. Not the least extraordinary part of his character was his indifference to the study of mechanics as a science: few men ever read more on general topics, and less on this—the knowledge of which of all others we might consider to have been essential to the proper development of his conceptions; and he never was at any period of his life a " scientific" man, in the present absurd acceptation of the term. It has been said that he never resolved an algebraic equation in his life,—like Ferguson, he got at the truth by geometrical methods; and at one period of his life it was his favourite amusement to represent by geometrical figures, various tables which he had sometimes occasion to consult in directing the proportions of his engines. These, it is said, gave his pupil Playfair the idea of his ingenious system of Linear Arithmetic. We have already given his own account of the origin of his invention being totally independent of any idea furnished by Dr Black, and in illustration we might notice the exquisite mechanism of the parallel motion. On being asked whether he could trace its invention to any previous chain of reasoning, he replied in the negative, adding, with his delightful candour, "that he was surprised himself at the perfection of its action; and in looking at it for the first time, he had all the pleasure of novelty which could have arisen had it been invented by another person." Indeed he was accustomed to consider those inventions which had carried his name to every corner of the world, as thoughts so obvious as must have occurred to thousands of others before his time, and that he alone was more fortunate than they, only in being the first who had put them to the test of experiment. His inventions, if he valued them lightly, occurred without effort. "Whether he had ever been a dexterous operative mechanic we have now no means of learning, but he certainly never attempted to assist in making models, or putting any of his own plans into execution after he came to England, whatever he might have done at an earlier period of his life. He employed most of his time in drawing or writing letters, but very little of it in superintending the operations that were going on. This probably arose from his feeling that he thought and contrived to the best purpose when his mind was left entirely to itself; though, on the other hand, it had the disadvantage that much more time was taken in realizing his ideas than otherwise would have been. The house in which he lived, near Birmingham, was two miles from Soho, where all the work was carried on: to this he seldom went above once a week to see what was doing, and when he did go there he seldom staid half an hour. As for Mr Bolton he never took any part (about 1786) in the manufacturing of the engines; his time being completely occupied in arrangements for obtaining the confidence and approbation of the public, and in providing the means of extending the use of the engine." *Memoir* by Playfair.

140. Francis Jeffrey, Esq.
141. *Repertory of Arts*, vol. XL. page 194. Second Series.
142. *Description of Masterman's Patent Rotatory Steam Engine*, London, 1822.
143. *Description,* p.24.
144. It may be urged that when a patent is enrolled it is open for public inspection. In one sense it is so. Those who have time to spend in a forenoon, and who live in London, (for the thing is sealed to all intents and purposes to country mechanics who form nine-tenths of the whole number), and who can decipher the handwriting of a Chancery Record, and who can pay a high fee for the privilege of looking at it, may certainly read the specification of a patent, and inspect the drawings; but if, in the few hours he is allowed to inspect them, he should be unable to understand the drawings or descriptions, and wish to have copies of either the one or the other, he must be prepared to pay from ten to twenty or thirty guineas for each patent; if he should be, for instance, an inventor of some steam mechanism, and before running to the expense of from 300 to 400 pounds in law fees to take out a patent, he wishes to know (as all prudent men ought) what has already been enrolled in the patent office in his particular department, this information in his case, we will venture to say, could not be obtained for less a sum than between £1500 or £9000!!! To the Editors of the Repertory of Arts, Tilloch's Philosophical Journal, the *London Journal of Science* and Gill's *Technical Repository*, and the Monthly Magazines which describe some of the inventions for which patents are granted, the mechanical world is under the greatest obligations. But owing to the enormous expense of having copies of patents made by the clerks in the patent office, and the rigid enforcement of a rule by which persons who may have paid a high fee for the inspection of a patent, are prevented from using pen, ink, pencil or paper, for the purpose of making a memorandum even of a date; these meritorious publications do not contain a twentieth part of those patents which are periodically granted, and generally those only which are sent to the editors by the more liberal patentees themselves. We know of nothing that would confer a greater and more lasting benefit on the mechanical world, than some legislative enactment, directing that the specifications and drawings of all patents should *be printed at length*, as soon as they are granted. Until this is done, no, patentee can feel assured that the £300 or £400 he has expended in *law-fees* are not absolutely thrown away.

Also available from Nonsuch Publishing

Alexander, Boyd (ed.)	*The Journal of William Beckford in Portugal and Spain*	978 1 84588 010 1
Brontë, Rev. Patrick	*The Letters of the Rev. Patrick Brontë*	978 1 84588 066 8
Broughton, S.D.	*Letters from Portugal, Spain and France*	978 1 84588 030 9
Brunel, Isambard	*The Life of Isambard Kingdom Brunel, Civil Engineer*	978 1 84588 031 6
Coleman, E.C. (ed.)	*The Travels of Sir John Mandeville, 1322–1356*	978 184588 075 0
Corbett, Sir Julian	*The Campaign of Trafalgar*	978 1 84588 059 0
Duff, Charles	*A Handbook on Hanging*	978 1 84588 141 2
Eyre, Lt Vincent	*The Military Operations at Cabul*	978 1 84588 012 5
Fothergill, A. Brian	*Beckford of Fonthill*	978 1 84588 085 9
Fothergill, A. Brian	*Sir William Hamilton: Envoy Extraordinary*	978 1 84588 042 2
Gooch, Sir Daniel	*The Diaries of Sir Daniel Gooch*	978 1 84588 016 3
Greenwood, Lt John	*The Campaign in Afghanistan*	978 1 84588 004 0
Hammond, J.L. and Barbara	*The Village Labourer*	978 1 84588 056 9
Hawkes, Francis L.	*Commodore Perry and the Opening of Japan*	978 1 84588 026 2
Helps, Sir Arthur	*The Life and Labours of Thomas Brassey*	978 1 84588 011 8
Hill, Wg Cdr Roderic	*The Baghdad Air Mail*	978 1 84588 009 5
Hudson, W.H.	*Idle Days in Patagonia*	978 1 84588 024 8
Jefferies, Richard	*Wildlife in a Southern County*	978 1 84588 064 4
Livingstone, David and Charles	*Expedition to the Zambesi and its Tributaries*	978 1 84588 065 1
Matthews, Henry	*Diary of an Invalid*	978 1 84588 017 0
Park, Mungo	*Travels in the Interior of Africa*	978 1 84588 068 2
Scott, Capt. Robert F.	*The Voyage of the Discovery, Vol. One*	978 1 84588 057 6
Ségur, Gen. Count Philippe de	*Memoirs of an Aide de Camp of Napoleon, 1800–1812*	978 1 84588 005 7
Simmonds, P.L.	*Sir John Franklin and the Arctic Regions*	978 1 84588 007 1

For forthcoming titles and sales information see

www.nonsuch-publishing.com